# AQA Certifica[te] Geography (iGCSE)

**LEVEL 1/2**

Stephen Durman
Simon Ross

**OXFORD**
UNIVERSITY PRESS

# OXFORD
UNIVERSITY PRESS

Great Clarendon Street, Oxford, OX2 6DP, United Kingdom

Oxford University Press is a department of the University of Oxford.
It furthers the University's objective of excellence in research, scholarship,
and education by publishing worldwide. Oxford is a registered trade mark of
Oxford University Press in the UK and in certain other countries

First published by Nelson Thornes Ltd in 2013
This edition published by Oxford University Press in 2014

British Library Cataloguing in Publication Data
Data available

978-1-4085-2130-4

3

Printed in Spain

**Acknowledgements**

**Cover photograph:** Sava Alexandru/iStockphoto
**Illustrations:** David Russell Illustration, Tim Jay, Peters & Zabransky and
GreenGate Publishing Services
**Page make-up:** GreenGate Publishing Services

Although we have made every effort to trace and contact all
copyright holders before publication this has not been possible in all
cases. If notified, the publisher will rectify any errors or omissions at
the earliest opportunity.

# Contents

**4  Contents**

Geography plays a crucial role in understanding some of the most pressing challenges facing the world today. This course includes a study of our natural environments and the pressures they face. It also focuses on the links between different parts of the world and how human activity affects the environment. The course looks at how the world is changing globally and locally, and the choices that exist in managing our world for the future.

The Level 1/2 Certificate in Geography has three units; Unit 1 – Dynamic Physical World, Unit 2 – Global Human Issues and Unit 3 – Application of Geographical Skills and Decision Making. These will all be examined by written papers at the end of the course. Paper 3 is based on a theme taken from the subject content of Units 1 and 2.

## Unit 1 Dynamic Physical World

In the exam, you must answer two questions. This written paper represents 30 per cent of the total marks

### Topics

- Tectonic activity and hazards
- Ecosystems and global environments
- River processes and pressures
- Coastal processes and pressures

Questions will consist of short, structured responses as well as opportunities for extended writing.

These topics are intended to provide you with a solid foundation in physical geography to enable you to fully appreciate the physical world in which you live.

A great deal of emphasis has been placed on topicality, using up-to-date case studies and examples to explore themes and concepts, such as the 2011 Japanese tsunami, the 2010 eruption of Eyjafjallajökull in Iceland and recent floods in Cockermouth.

All of the physical geography topics have a link to human activities. For example, in *Tectonic activity and hazards* you will study the impact of tectonic hazards on people's lives. The impacts of flooding in rich and poor countries are discussed in *River processes and pressures*. Sustainable development is a key concept in all topics. For example, in *Coastal processes and pressures*, strategies for coastal defence and coastal zone management are underpinned by the need for a sustainable approach. In *Ecosystems and global environments*, sustainability is at the heart of ecosystem management, whether it is hot deserts or tropical rainforest.

## Unit 2 Global Human Issues

In the exam, you must answer two questions. This written paper represents 30 per cent of the total marks

### Topics

- Contemporary population issues
- Contemporary issues in urban settlements
- Globalisation in the contemporary world
- Contemporary issues in tourism

Questions will be a mix of short, structured responses as well as opportunities for extended writing.

These topics have been chosen to reflect current thinking and interests in geography in the 21st century. They all involve the study of recent issues and extensive use is made of up-to-date case studies and examples. For example, in *Contemporary population issues*, a detailed study is made of the implications of China's controversial 'one child policy'. The current (and future) issue of ageing populations in Europe is also considered. Sustainable development is a significant aspect of human geography. For example, in *Contemporary issues in tourism*, you will discuss the issues surrounding mass tourism as well as examining recent trends such as ecotourism and adventure tourism. In *Globalisation in the contemporary world*, topics such as renewable energy and pollution control are also tackled. As you study human geography there will be many opportunities for you to discuss and debate real world issues, such as traffic congestion and its management in *Contemporary issues in urban settlements* and global warming in *Globalisation in the contemporary world*. You will be able to formulate and share your own opinions and, hopefully, you will be able to make a positive difference to the world around you.

# 1 Tectonic activity and hazards

## 1.1 Why is the earth's crust unstable?

### The structure of the earth

The **crust** – the outer layer of the earth – is relatively thin (diagram **A**). The crust is not one single piece of skin, like that of an apple. Instead, it is split into **plates** of varying size and at **plate margins** it is liable to move. This is because the slabs of crust float on the semi-molten upper **mantle**. **Convection currents** within the mantle determine the direction of plate movement. Therefore, in some cases the plates are moving together and sometimes they are moving apart. There are two types of crust: oceanic and continental (diagram **B**). The location of the plates, plate boundaries and direction of movement of the plates is shown in map **C**.

**In this section you will learn**

about the structure of the earth and the difference between oceanic and continental crust

how and why destructive, constructive and conservative plate margins are different

how the earth's crust is unstable, especially at plate margins.

**Key terms**

**Crust**: the outer layer of the earth.

**Plate**: a section of the earth's crust.

**Plate margin**: the boundary where two plates meet.

**Mantle**: the dense, mostly solid layer of the earth between the outer core and the crust.

**Convection currents**: the circular currents of heat in the mantle.

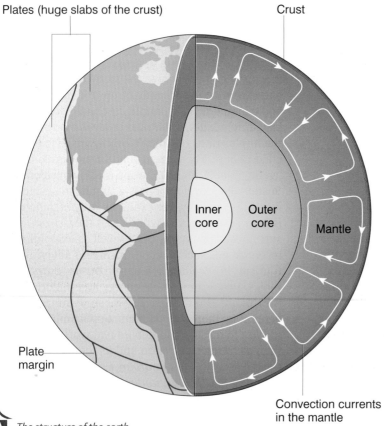

Plates (huge slabs of the crust)

Crust

Inner core

Outer core

Mantle

Plate margin

Convection currents in the mantle

**A** *The structure of the earth*

**Did you know** ???????

The earth's crust is divided into 14 major plates and 38 minor plates, making a total of 52 jigsaw pieces altogether.

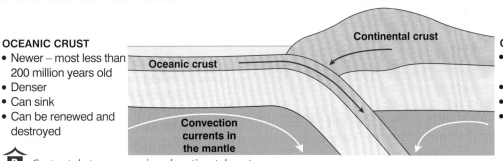

**OCEANIC CRUST**
- Newer – most less than 200 million years old
- Denser
- Can sink
- Can be renewed and destroyed

**CONTINENTAL CRUST**
- Older – most over 1500 million years old
- Less dense
- Cannot sink
- Cannot be renewed or destroyed

**B** *Contrasts between oceanic and continental crust*

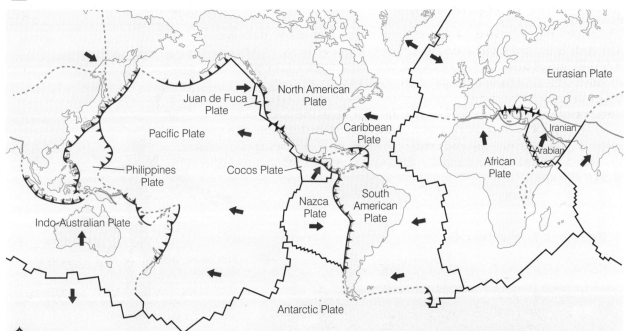

**C** *World tectonic plates and margins*

Map key

➡ Direction of plate movement

Destructive margin – one plate sinks under another (subduction)

Destructive (collision) margin – two continental plates move together

Constructive margin – two plates move away from each other

Conservative margin – two plates slide alongside each other

Uncertain plate boundary

## Activities

1. Study diagram **A**.
   a. Draw a cross-section through the earth and label it to show the four layers.
   b. On your diagram, add a definition of each term you have labelled.

2. Study diagram **B**.
   a. Draw a simple labelled diagram to describe the differences between the two types of crust.
   b. Explain what the convection currents are in the mantle and how they cause plate movement.

3. Study map **C**. Copy and complete the table below.

| Plate margin | Direction of plate movement | Example of plate margin |
|---|---|---|
| Destructive – subduction | | |
| Destructive – collision | | |
| Constructive | | |
| Conservative | | |

# Types of plate margin

## Destructive plate margins

Convection currents in the mantle cause the plates to move together. If one plate is made from oceanic crust and the other from continental crust, the denser oceanic crust sinks under the lighter continental crust in a process known as **subduction**. Great pressure is exerted and the oceanic crust is destroyed as it melts to form magma.

If two **continental plates** meet each other, they collide rather than one sinking beneath the other. This **collision** boundary is a different type of destructive margin.

**Key terms**

**Destructive plate margin:** a plate margin where two plates are moving towards each other resulting in one plate sinking beneath the other.

**Constructive plate margin:** a plate margin where two plates are moving apart.

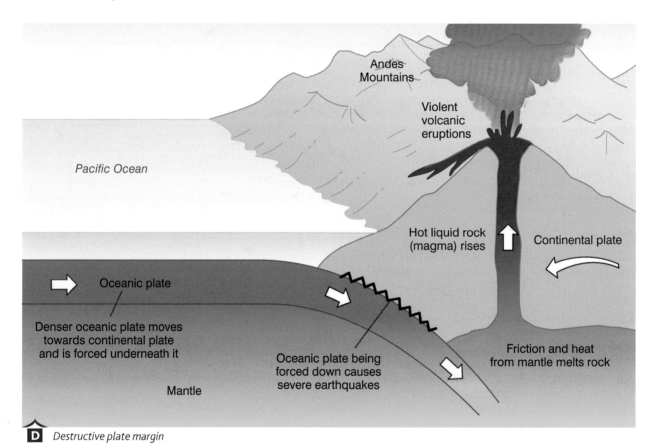

Andes Mountains

Violent volcanic eruptions

Pacific Ocean

Hot liquid rock (magma) rises

Continental plate

Oceanic plate

Denser oceanic plate moves towards continental plate and is forced underneath it

Oceanic plate being forced down causes severe earthquakes

Friction and heat from mantle melts rock

Mantle

**D**   *Destructive plate margin*

## Constructive plate margins

When plates move apart, a constructive plate boundary results. This usually happens under the oceans, as shown in map **C**. As these **oceanic plates** pull away from each other, cracks and fractures form between the plates where there is no solid crust. Magma forces its way into the cracks and makes its way to the surface to form **volcanoes**. In this way new land is formed as the plates gradually pull apart.

**Key terms**

**Continental plate:** a tectonic plate made of low density continental rock that will not sink under another plate.

**Oceanic plate:** a tectonic plate made of dense iron-rich rock that forms the ocean floor.

**Subduction:** when oceanic crust sinks under continental crust at a destructive margin.

**Collision:** when two plates of continental crust meet 'head on' and buckle.

**Volcano:** an opening in the earth's crust through which molten lava, ash and gases are ejected.

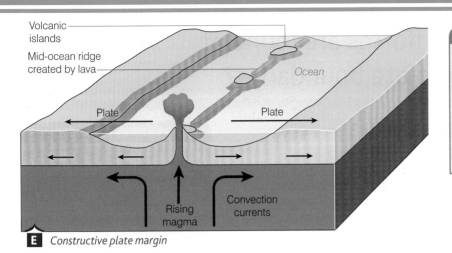

**E**  *Constructive plate margin*

## Conservative plate margins

At conservative plate margins, the plates are sliding past each other. They are moving in a similar (though not the same) direction, at slightly different angles and speeds. As one plate is moving faster than the other and in a slightly different direction, they tend to get stuck. Eventually, the build-up of pressure causes them to be released. This sudden release of pressure causes an **earthquake**. At a conservative margin, crust is being neither destroyed nor made. With no source of magma, volcanoes are absent.

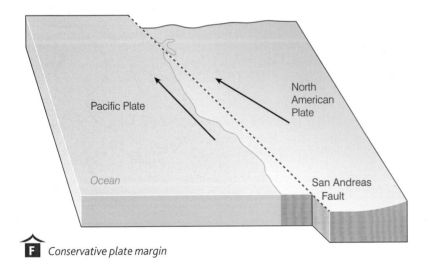

**F**  *Conservative plate margin*

# 1.2 What landforms are found at different plate boundaries?

## Fold mountains and ocean trenches

Young **fold mountains** (those that have been formed in the last 65 million years) are the highest areas in the world. All peaks over 7,000 m are in central Asia, including Mt Everest at 8,850 m. This dwarfs the highest mountain in England – Scafell Pike at 978 m. Young fold mountains include ranges such as the Himalayas, the Rockies, the Andes and the Alps. Older ranges of fold mountains that are less high due to erosion include the Cambrian mountains and the Cumbrian mountains in the UK. **Ocean trenches** form some of the deepest parts of the ocean. The distribution of young fold mountains and ocean trenches is shown in map **A**. Look back at map **C** on page 7 to see the link between these landforms and plate margins.

Look back at map **C** on page 7

> **In this section you will learn**
>
> why fold mountains and ocean trenches form at destructive plate margins
>
> the differences between composite volcanoes, which are associated with destructive plate margins, and shield volcanoes, which are associated with constructive plate margins.

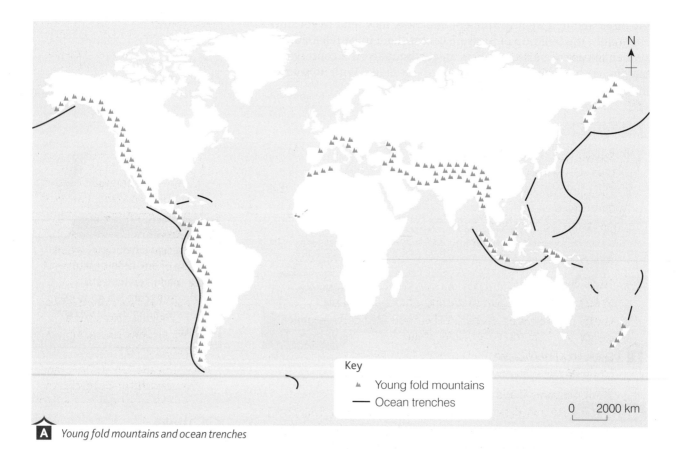

**Key**

- ▲ Young fold mountains
- — Ocean trenches

0    2000 km

**A** *Young fold mountains and ocean trenches*

Both fold mountains and ocean trenches result from plates moving together. If both landforms occur in the same area, they are found in association with subduction. If fold mountains occur by themselves, they are in areas where collision is taking place. Either way, the sequence relating to their formation is similar, as shown in diagram **B**.

> **Study tip**
>
> Make sure you know the correct sequence of formation of fold mountains.

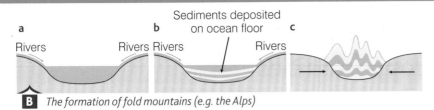

**B** *The formation of fold mountains (e.g. the Alps)*

a · Rivers · Rivers  b · Rivers · Sediments deposited on ocean floor · Rivers  c

## Composite and shield volcanoes

There are two types of volcano: **composite volcanoes**, which occur at destructive plate margins, and **shield volcanoes**, which occur at constructive plate margins (diagrams **C** and **D**).

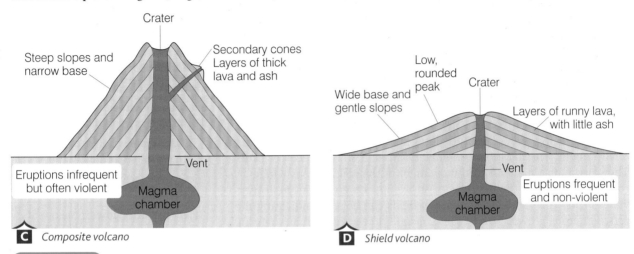

**C** *Composite volcano*

Crater
Steep slopes and narrow base
Secondary cones
Layers of thick lava and ash
Vent
Eruptions infrequent but often violent
Magma chamber

**D** *Shield volcano*

Low, rounded peak
Crater
Wide base and gentle slopes
Layers of runny lava, with little ash
Vent
Magma chamber
Eruptions frequent and non-violent

**OO links**

Investigate further facts and figures about landforms resulting from plate movements at **www.geology. com** and **www.extremescience.com**.

## Activities

1 Study map **C** on page 7 and map **A** opposite. You will need a world map outline showing the plates and plate boundaries.

a On your map, add the young fold mountains shown in map **A** and name them.

b Add the ocean trenches.

c At what type of plate boundary do fold mountains and ocean trenches form?

2 Study diagram **B**.

a Make a copy of each part of the diagram.

b Use the key terms on the right to help you label each diagram to explain the formation of fold mountains.

c Draw another diagram to explain the formation of ocean trenches (see diagram **D** on page 8).

3 Study diagrams **C** and **D**.

a On your own, write five questions you would ask to find out about the contrasts between a composite volcano and a shield volcano.

b Swap questions with a partner and answer their questions.

c What were the good points about the questions you have just answered? How might they be improved?

d Use an internet search engine to find images of composite and shield volcanoes. Draw a sketch of each image and label them to show the characteristics of each type of volcano.

e Summarise how and why composite and shield volcanoes are different.

## Key terms

**Fold mountains**: large mountain ranges where rock layers have been crumpled as they have been forced together.

**Ocean trenches**: deep sections of the ocean, usually where an oceanic plate is sinking below a continental plate.

**Composite volcano**: a steep-sided volcano that is made up of a variety of materials, such as lava and ash.

**Shield volcano**: a broad volcano that is mostly made up of lava.

# 1.3   How do tectonic landscapes provide economic opportunities?

## ■ The Andes and its people

The Andes is a range of young fold mountains; a tectonic landscape formed some 65 million years ago (map **A**, page 10). It is the longest range of fold mountains in the world at 7,000 km and extends the length of South America. The Andes are about 300 km in width and have an average height of 4,000 m.

## Farming

Despite the high altitude of the Andes, the mountain slopes are used for farming. In Bolivia, many **subsistence** farmers grow a variety of crops on the steep slopes, including potatoes which are a main source of food. The use of **terraces** (photo **A**) creates areas of flat land on the slopes. Terracing offers other advantages in trying to farm in this harsh environment. The flat areas retain water in an area that receives little. They also limit the downward movement of the soil in areas where the soils are thin in the first place. Most crops are grown in the lower valleys (photo **B**) and a patchwork of fields can be seen indicating the range of crops grown. Some cash crops are produced such as soybeans, rice and cotton.

Llamas are synonymous with the Andes. For hundreds of years they have been pack animals, carrying materials for **irrigation** and buildings into inhospitable and inaccessible areas. The ancient settlement of Machu Picchu (photo **C**) relied on llamas to transport materials and goods due to its remote location. These surefooted animals can carry over 25 per cent of their body weight (125–200 kg). The mining industry often relied on them as a form of transport. Today, this use still exists – largely of male llamas. The females are used for meat and milk, and their wool is used in clothes as well as rugs.

### In this section you will learn

how people use tectonic landscapes

how people adapt to the difficult conditions within them.

### Key terms

**Subsistence**: farming to provide food and other resources for the farmer's own family.

**Terraces**: steps cut into hillsides to create areas of flat land.

**Irrigation**: artificial watering of the land.

**Hydroelectric power**: the use of flowing water to turn turbines to generate electricity.

### Study tip

With reference to a specific range of fold mountains, ensure that you can list and explain the uses that people make of this environment.

**A**   *Farming in the Andes*

**B**   *The lower valleys*

## Mining

The Andes has a range of important minerals and the Andean countries rank in the top 10 for tin (Peru and Bolivia), nickel (Colombia), silver (Peru and Chile) and gold (Peru). More than half of Peru's exports are from mining. The Yanacocha gold mine (map **D** and photo **E**) is the largest gold mine in the world. It is a joint venture between a Peruvian mining company and a US-based one that has a 51 per cent share. It is an open pit and the gold-bearing rock is loosened by daily dynamite blasts. The rock is then sprayed with cyanide and the gold extracted from the resulting solution. This can lead to contamination of water supplies.

**C**   *Machu Picchu*

The nearby town of Cajamarca has grown from 30,000 inhabitants (when the mine began) to about 300,000 in 2010. This brings with it alternative sources of jobs. However, this growth also brings many problems, including a lack of services and an increased crime rate.

## Hydroelectric power

The steep slopes and narrow valleys that limit farming are an advantage for **hydroelectric power** (HEP). They can be more easily dammed than wider valleys and the relief encourages the rapid fall (flow) of water needed to ensure the generation of electricity. The melting snow in spring increases the supply of water, but the variation throughout the year is a disadvantage rather than an advantage. In 2009 the El Platanal HEP power plant began to generate electricity. Involving a huge dam across the Cânete River, the $US200 million project is the second largest in Peru (map **D**).

**D**   *The location of Yanacocha gold mine and the El Platanal HEP project*

## Tourism

There are many natural attractions in the Andes such as mountain peaks, volcanoes, glaciers and lakes. Some tourist attractions show how people settled in these inhospitable areas, such as the remains of early settlements built by the Incas like Machu Picchu (photo **C**). The Inca Trail combines both (extract **F**).

**E**   *Yanacocha gold mine*

# The Inca Trail is South America's best-known trek

The Inca Trail is South America's best-known trek. It combines a stunning mix of Inca ruins, mountain scenery, lush forest and tropical jungle. Over 250 species of orchid exist in the Machu Picchu historic sanctuary, as well as numerous species of birds. The sanctuary is an important natural and archaeological reserve – it is only one of 23 UNESCO world heritage sites to be classified as important both culturally and naturally.

The Inca Trail is a hike that finishes at Machu Picchu, the sacred mysterious 'Lost City of the Incas'. The 45 km trek is usually covered in four days, arriving at Machu Picchu at daybreak on the final day before returning to Cusco by train in the afternoon. The trek is best undertaken from April to October, when the weather is drier. Any fit person should be able to cover the route. It is fairly challenging and altitudes of 4,200 m are reached, so it is important to be well acclimatised.

**F**  *The Inca Trail*

## ∞ links

Investigate further facts and discussion about gold mining by searching on **www.google.co.uk**.

For more on the Inca Trail, go to **www.incatrailperu.com**.

## Activities

1  Study map **C** on page 7 and an atlas map of South America. Locate the Andes on the map of South America. Using these resources, produce a fact file about the Andes to provide the following information:

- location
- countries
- plates responsible for mountains
- direction of plate movement
- type of plate margin
- dimensions
- highest peak and its location.

2  Study photos **A** and **B**.

a  Describe how the mountains are used for crops.

b  Outline the role of the llama in the life of Andean residents.

c  Explain how difficulties of steep relief and poor soils are overcome.

3  Study map **D**.

a  Work in pairs to produce a short presentation to include:

- a map showing the location of the Yanacocha gold mine
- the advantages of the area for the mine and for Peru as a whole
- the disadvantages of the area for the mine and for Peru as a whole.

b  Show your presentation to another group.

4  Research HEP and specifically that produced in the Andes. On one side of A4 paper, define HEP, explain why the Andes is suitable for the development of HEP and give details of one HEP project. Include at least one map and one photo to illustrate your information.

5  Read extract **F**. Imagine you have visited Peru and part of your holiday included the Inca Trail. Write a postcard to send home, which includes the following:

- a map showing the trail
- a description of how you got to the trail
- a description of the attractions (natural and human) that you saw. Explain why these were so awesome and different to what you had seen previously.

# 1.4   What are the different types of volcanic activity?

Volcanoes form when the earth's plates pull apart along constructive plate margins, allowing magma to rise to the surface to create a shield volcano. They can also occur where one plate is subducted or dives below another along a destructive plate margin. The oceanic plate descends under the continental plate because it is denser. As the plate descends it starts to melt due to the friction caused by the movement between the plates. This melted plate is now hot, liquid rock (magma). The magma rises along cracks and faults in the continental plate. If it reaches the surface, the liquid rock forms a composite volcano.

About five per cent of volcanoes do not take place near the margins of tectonic plates. They are found over very hot places in the earth's interior called hot spots. **Hot spots** are created by mantle plumes, hot currents that rise all the way from the core through the mantle. When mantle plumes come up under the crust, they burn their way through to become hot spot volcanoes. Famous hot spot volcanoes include the Hawaiian island volcanoes (map **B**). As the Pacific plate slowly moved over the hot spot, the islands in the Hawaiian chain were built one at a time by volcanic eruptions (diagram **A**). Island formation is still happening on Hawaii every time the shield volcanoes Kilauea and Mauna Loa erupt. Hot spot volcanoes ooze runny lava that spreads out to create shield volcanoes.

**In this section you will learn**

about the causes of volcanic activity

about some of the features found close to volcanoes.

**Key term**

Hot spot: a section of the earth's crust where plumes of magma rise, weakening the crust. These are away from plate boundaries.

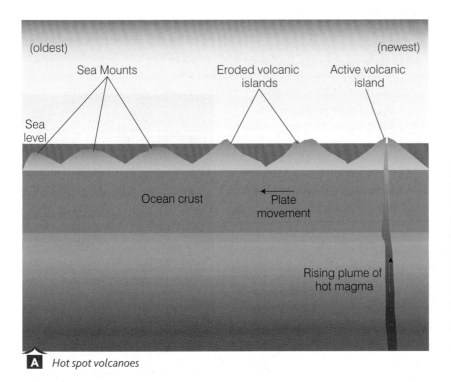

**A**  Hot spot volcanoes

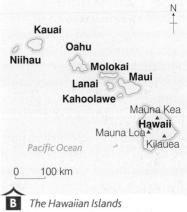

**B**  The Hawaiian Islands

Hot springs and geysers form when cold groundwater seeps down and is heated by the rocks touching the underlying magma chamber. The hot water then rises to the surface.

Hot springs occur when this heated water forms a pool on the surface of the Earth. Hot springs vary in temperature and can be calm, bubbly, or boiling depending on how hot the magma chamber is. When the hot water travels up, it dissolves material from the surrounding bedrock and brings this material up to the surface with it.

**C**    *Hot spring pond, Niseko, Hokkaido, Japan*

A geyser is a type of hot spring that periodically erupts, shooting columns of water and steam into the air. In contrast to hot springs, where the heated water has a simple path to travel upwards to reach the surface, geysers have a complex network of underground tunnels and reservoirs that trap the water and delay its arrival to the surface. In a restricted channel, steam bubbles generated by the rising hot water build up behind the constrictions. When the build-up reaches a critical level, the steam squeezes through the narrow passageways, forcing the surface water to overflow. The easing of the pressure at the surface causes a sudden drop in pressure in the superheated water in the underground chamber. A violent chain reaction is started, resulting in huge steam explosions in which the rising, boiling water expands 1,500 times or more by volume. This expanding body of superheated water is thrown out through the narrow channel in a high-pressure fountain-burst, or geyser. This keeps happening to all the water within the chambers underground until there is no longer enough water left to continue the eruption. Groundwater then starts seeping back into this underground network, starting the cycle all over again.

Many of these geothermal features are very colourful. These colours are due to the substances found in the water, and the colour is a very good indicator of what these substances are. If a spring has a red colour to it, most likely it is caused by a large amount of iron. If it is yellow, it is probably due to the presence of sulphur, often associated with the smell of rotten eggs. Pinks and whites are often caused by the presence of calcium.

## Key terms

**Hot spring:** A spring of naturally hot water, usually heated by underground volcanic activity.

**Geyser:** a geothermal feature in which water erupts into the air under pressure.

## Did you know ??????

Hot springs in volcanic areas are often at or near the boiling point. People have been seriously burned and even killed by entering these springs.

## Did you know ??????

Because heated water can hold more dissolved solids, hot springs often have a very high mineral content, containing everything from simple calcium to lithium, and even radium. Because of the claimed medical value some of these springs have, they are often popular tourist destinations, and locations of special clinics for people with disabilities.

**D**    *Pohutu and Prince of Wales geysers, Rotarua, New Zealand*

# 1.5   How do volcanoes affect people?

## The distribution of volcanoes

Volcanoes are an example of a **natural hazard**. Their spread relates closely to plate margins (map **A** below and map **C**, page 7). The area around the Pacific Ocean is especially prone to volcanoes and is known as 'the Pacific Ring of Fire'. Occasionally, active volcanoes are found away from plate margins. We are going to study two recent volcanic eruptions that have had significant impacts on people's lives.

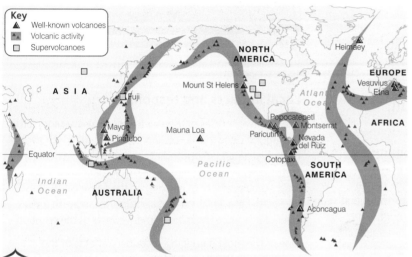

**A**   *The distribution of active volcanoes and supervolcanoes*

### Case study

## The eruption of Nyiragongo, Africa (2002)

On 17 January 2002, Nyiragongo volcano in the Democratic Republic of Congo was disturbed by the movement of plates along the East African rift valley. This led to lava spilling southwards in three streams. The speed of the lava reached 60 kph, which is especially fast. The lava flowed across the runway at Goma airport (photo **C**) and through the town, splitting it in half (map **B**). The lava destroyed many homes as well as roads and water pipes, set off explosions in fuel stores and power plants, and killed 45 people. These were the **primary effects**. In addition, there were many **secondary effects**. Half a million people fled from Goma into neighbouring Rwanda to escape the lava. They spent the night sleeping on the streets of Gisenyi. Here, there was no shelter, electricity or clean water as the area could not cope with the influx. Diseases such as cholera were a real risk. People were frightened of going back. However, looting was a problem in Goma and many residents returned within a week in the hope of receiving **aid**. In the aftermath of the eruption, water had to be supplied in tankers. Aid agencies, including Christian Aid and Oxfam, were involved in the distribution of food, medicine and blankets.

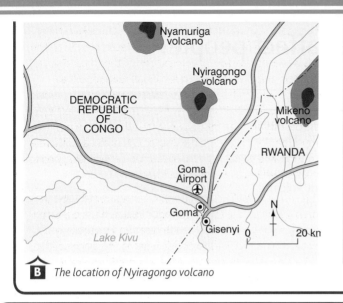

**B**    The location of Nyiragongo volcano

**C**    Lava on the runway at Goma airport

# The eruption of Eyjafjallajökull, Iceland 2010

Iceland lies on the Mid-Atlantic Ridge, a constructive plate margin separating the Eurasian plate from the North American plate (map **C** page 7). As the plates move apart magma rises to the surface to form several active volcanoes located in a belt running roughly SW–NE through the centre of Iceland (map **D**). Eyjafjallajökull (1,666 m high) is located beneath an ice cap in southern Iceland 125 km south east of the capital Reykjavik.

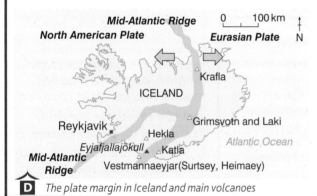

**D**    The plate margin in Iceland and main volcanoes

## The eruption

In March 2010, magma broke through the crust beneath Eyjafjallajökull glacier. This was the start of two months of dramatic and powerful eruptions that would have an impact on people across the globe. The eruptions in March were mostly lava eruptions. Whilst they were spectacular and fiery they represented little threat to local communities.

However, on 14 April a new phase began which was much more explosive. Over a period of several days in mid-April violent eruptions belched huge quantities of ash into the atmosphere.

## Local impacts and responses

The heavier particles of ash (such as black gritty sand) fell to the ground close to the volcano, forcing hundreds of people to be evacuated (an **immediate response**) from their farms and villages. As day turned to night, rescuers wore face masks to prevent them choking on the dense cloud of ash. These ash falls, which coated agricultural land with a thick layer of ash, were the main primary effects of the eruption.

One of the most damaging secondary effects of the eruption was flooding. As the eruption occurred beneath a glacier, a huge amount of meltwater was produced. Vast torrents of water flowed out from under the ice. Sections of embankment that supported the main highway in southern Iceland were deliberately breached by the authorities to allow the floodwaters to pass through to the sea. This action successfully prevented expensive bridges being destroyed. After the eruption, bulldozers were quickly able to rebuild the embankments and within a few weeks the highway was reconstructed.

## International impacts and responses

The eruption of Eyjafjallajökull became an international event in mid-April 2010 as the cloud of fine ash spread south-eastwards towards the rest of Europe (map **E**). Concerned about the possible harmful effects of ash on aeroplane jet engines, large sections of European airspace closed down. Passenger and freight traffic throughout much of Europe ground to a halt.

**Key terms**

**Immediate responses**: how people react during a disaster and straight afterwards.

**Long-term responses**: later reactions that happen in the weeks, months and years after the event.

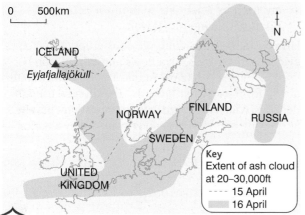

**E** *Map showing the spread of the ash cloud over Europe*

The knock-on effects were extensive and were felt across the world (table **F**). Business people and tourists were stranded, unable to travel in to or out of Western Europe. Industrial production was affected as raw materials could not be flown in and products could not be exported by air. As far away as Kenya, farm workers lost their jobs or suffered pay cuts as fresh produce such as flowers and beans perished, unable to be flown to European supermarkets. The airline companies and airport operators lost huge amounts of money.

Some people felt that the closures were an over-reaction and that aeroplanes could fly safely through low concentrations of ash. However, a scientific review conducted after the eruption concluded that under the circumstances it had been right to close the airspace. Further research will be carried out as a **long-term response** to find better ways of monitoring ash concentrations and improving forecast models.

## Prediction, planning and preparation

The Icelandic Meteorological Office monitored earth movements and emissions of volcanic material, and issued warnings about the potential eruption. Earthquake activity had been significantly increasing since December 2009, and this was accompanied by deformation of the volcano since the start of 2010. Prior to the ash cloud reaching UK air space on April 15th the volcano had been erupting for almost four weeks. The IMO's weather radar on the southwest tip of the country showed the height of the ash plume, which was important for calculating the distribution of the ash. There is a nationwide digital network of seismic stations, designed to forecast earthquake and volcanic activity.

The London VAAC (Volcanic Ash Advisory Centre) also gave information on the ash plume and predicted the expected changes every 24 hours. The information was then used by aviation authorities to decide whether airspace needed to be closed to prevent aircraft encountering volcanic ash. This event was therefore tracked and prepared for, and the ash cloud was tracked by satellite by many countries.

## Not all bad news!

There can be positive impacts of volcanic activity. Eyjafjallajökull has become a new Icelandic tourist attraction with its own visitor centre. The Icelandic Tourist Board now encourages people to visit the 'land of ice and fire' bringing benefits to local people and industries.

Lava and ash are rich in nutrients making the soils in volcanic areas very fertile and good for agricultural use. The rocks themselves can be used for building. The valuable source of geothermal energy just below the surface of the earth is usually relatively easily to access. This can be used to produce electricity and provide hot water for industrial processes, heating swimming pools and even melting snow on pavements.

Small quantities of valuable minerals such as copper, gold, silver, lead, and zinc are often found in volcanic regions. Deposits of sulphur can be mined from volcanic craters and used to bleach sugar and make matches and fertiliser.

**F** *Impacts of the eruption*

| Local | National | International |
|---|---|---|
| • Some 800 local people living close to the volcano had to be evacuated. | • Tourism was affected with the number of international tourists dropping during the summer 2010. This affected the Icelandic economy as well as local people's jobs and incomes. | • Over a period of 8 days, some 100,000 flights were cancelled accounting for 48 per cent of total air traffic. Ten million passengers were affected. The Airport Operators' Association estimated the total loses to be £80 million. |
| • Homes and roads were damaged and services (electricity, water, etc) disrupted. | | • Industries were affected by a lack of imported raw materials, e.g. Honda announced a partial halt to production. |
| • Local flood defences had to be reconstructed. | • Road transport was disrupted as roads were washed away by floods. | • Fresh food could not be imported, affecting supermarkets and producers across the world. |
| • Crops (particularly grass used for hay) were damaged by heavy falls of ash. | • Agricultural production was affected as crops were smothered by a thick layer of ash. | • In sport, a number of international events were affected by the flight ban including the Japanese Motorcycle Grand Prix, French rugby league teams competing in the Challenge Cup and the Boston Marathon on 19 April. |
| • Local water supplies were contaminated with fluoride. | • Reconstruction of roads and services was expensive. | |

## Monitoring and predicting volcanoes

Unlike earthquakes, volcanoes usually provide warning signs of an impending eruption. Active volcanoes across the world are monitored by scientists using high technology equipment and warnings are issued if a volcano seems likely to erupt.

- An increase in earthquake activity is an indication that magma is rising beneath a volcano, causing rocks to crack and fracture. Map **G** shows the concentration of earthquakes just before the eruption of Eyjafjallajökull in 2010. Earthquakes are recorded using instruments called seismometers that are placed in the ground.

- Tiltmeters can be placed on the ground to measure slight changes in the tilt of the ground caused by rising magma (diagram **H**). Just before Mt St Helens erupted in 1980 the north face of the volcano bulged as magma rose within the volcano.

- Digital cameras (photo **I**) can be placed on the rim of craters to record small eruptions or landslides that might indicate rising magma (e.g. White Island, New Zealand). Controlled from a distance, this is a safe way to monitor crater activity.

- Global positioning systems (GPS) use satellite technology to measure very slight changes of as little as 1 mm (photo **J**). Laser beams can also be used to measure changes in distance between two fixed points on a volcano. If the volcano swells, the distance between two places will increase.

- Gases emitted from a volcano, such as sulphur dioxide, can change in concentration prior to an eruption. Continuous gas monitoring stations are used by scientists to monitor activity at Kilauea volcano on Hawaii.

Historic information from previous eruptions, such as evidence of ash falls, lava flows and **lahars**, can be used to construct **hazard maps**. These identify zones at risk from particular hazards. They can be used in deciding which areas are safe for developments such as housing and also in making plans for evacuations.

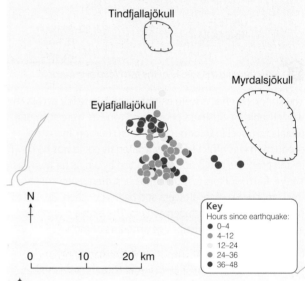

**G**    *Earthquakes recorded over 48 hrs prior to 3 March 2010 indicating that an eruption is likely at Eyjafjallajökull*

> **Key terms**
>
> **Lahar**: mudflows resulting from ash mixing with melting ice or water – a secondary effect of a volcano.
>
> **Hazard maps**: a map that shows areas that are at risk from hazards such as earthquakes, volcanoes, landslides, floods and tsunamis.

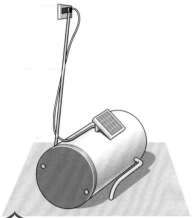

**H**    *Tiltmeters detect a change in slope caused by shifting magma beneath the surface*

**I**    *Time-lapse cameras in the crater allow geologists to make safe observations*

**J**   *GPS use satellites to detect minute movement*

∞ **links**

Further information with many superb links about the eruption of Eyjafjallajökull can be found at **www. geographyinthenews.rgs.org** search for Eyjafjallajökull.

## Activities

1   Study map **A** on page 17.

a   Describe the distribution of the world's active volcanoes.

b   Compare the distribution with map **C** on page 7. Provide evidence to support the statement that 'volcanoes usually occur at plate boundaries'.

c   Explain the cause of the Nyiragongo eruption.

2   Study map **B** and photo **C** on page 18.

a   Draw a sketch map based on the map to show the location of the volcano, the countries and the main towns. Use an arrow to show the flow of the lava towards Goma.

b   Label your sketch map to show the primary and secondary effects of the eruption at Nyiragongo.

3   Read the text on page 18. Produce a front-page newspaper report for 15 April 2010 detailing the eruption of Eyjafjallajökull. Your article should be illustrated with a map showing the location of the volcano and the tectonic plates and at least one labelled photo. The text should describe the nature of the eruption and its cause. Refer to the tectonic details and suggest the likely impacts of the eruption.

4   Study the text on pages 18 and 19 and Table **F**.

a   How were local farmers and their families affected by the eruption of Eyjafjallajökull?

b   Describe the impact of the eruption on tourism.

c   How did the closure of European airspace affect

■   European businesses and industries

■   supermarkets

■   farming in Kenya

■   sporting events?

5   Read the text in this section.

a   Describe and compare the immediate responses to the eruptions at Nyiragongo and Eyjafjallajökull.

b   Describe the long-term responses at Eyjafjallajökull.

6   Tourism is one positive effect of volcanoes. Use the internet to research answers to the following questions.

a   How has Eyjafjallajökull become a tourist attraction?

b   Research two other positive effects of volcanic activity in Iceland. Present your information to describe each positive effect and provide an example of where this occurs.

7   Study the text on page 20 together with map **G**, diagram **H**, photos **I** and **J**.

a   Make a list of the different means of monitoring and predicting volcanoes.

b   Describe in detail two methods of monitoring volcanoes. Explain the reasons for each form of monitoring and how it indicates that an eruption is likely.

c   How successful do you think these methods are in predicting volcanoes?

# 1.6    What is a supervolcano?

A **supervolcano** is a huge volcano with the potential to erupt catastrophically and with enormous power. Look back to map **A** on page 17 to locate the world's supervolcanoes.

## ◼ Characteristics of a supervolcano

Supervolcanoes are on a much bigger scale than volcanoes. They emit at least 1,000 km³ of material – compare this with an eruption on the magnitude of Mt St Helens in 1980, which emitted some 1 km³. Supervolcanoes do not look like a volcano with its characteristic cones. Instead, they are large depressions called **calderas**, often marked by a rim of higher land around the edges (diagram **A**).

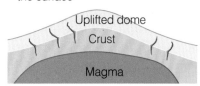

1 Rising magma cannot escape, and a large bulge appears on the surface

Uplifted dome
Crust
Magma

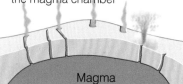

2 Cracks appear in the surface and gas and ash erupt from the magma chamber

Magma

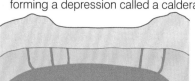

3 The magma chamber collapses, forming a depression called a caldera

 **A**   *The formation of a supervolcano*

## ◼ The Yellowstone supervolcano

Many visitors stand in awe looking at Old Faithful and the **geothermal** features of the Norris Geyser basin without realising the vulnerability of where they are standing. The very forces that created such a unique area could be responsible for its destruction and threaten the existence of people – at the very least in North America, if not globally. Map **B** shows the area of Yellowstone and some of its attractions.

There is evidence that the magma beneath Yellowstone is shifting. The caldera is bulging up beneath Lake Yellowstone. There are signs of increasing activity at Norris and the ground has risen 70 cm in places. Is this just part of a natural cycle? The magma chamber beneath Yellowstone is believed to be 80 km long, 40 km wide and 8 km deep. It is not known whether the magma is on top of other materials, which would be necessary for an eruption.

Eruptions have occurred at this hot spot 2 million years ago, 1.3 million years ago and 630,000 years ago. An eruption today would have a catastrophic effect. It is potentially five times the minimum size for a supervolcanic eruption by the size of the magma chamber.

### In this section you will learn

what a supervolcano is and how it differs from a volcano

the potential impact of a supervolcano eruption in contrast to a volcano eruption.

### Key terms

**Supervolcano**: a mega colossal volcano that erupts at least 1,000 km3 of material.

**Caldera**: the depression of the supervolcano marking the collapsed magma chamber.

**Geothermal**: water that is heated beneath the ground, which comes to the surface in a variety of ways.

### Did you know ??????

The last supervolcano eruption occurred on Sumatra when Toba erupted 74,000 years ago. It is thought world temperatures fell by between 3 and 5°C. Toba exuded 3,000 km³ of magma and covered India in a layer of ash 15 cm deep.

### Study tip

Make notes to reinforce your understanding of the differences between a volcano and a supervolcano.

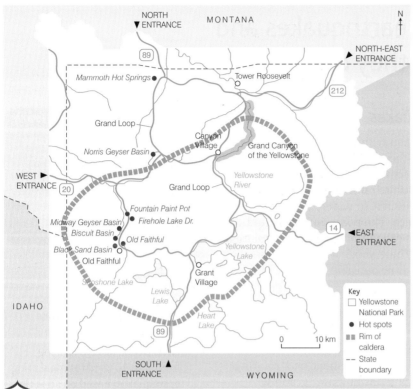

**B** *The location of the Yellowstone caldera*

An eruption is likely to destroy 10,000 km² of land, kill 87,000 people, 15 cm of ash would cover buildings within 1,000 km and 1 in 3 people affected would die. The ash would affect transport, electricity, water and farming. Lahars (mud flows) are a probability. The UK would await the arrival of the ash some five days later. Global climates would change, crops would fail and many people would die.

**C** *Cascade Geyser at Yellowstone National Park*

links

Investigate the Yellowstone supervolcano further at **www.discovery.com** and **www.bbc.co.uk**. Enter Yellowstone supervolcano into the search box in both websites.

# 1.7    What are earthquakes and where do they occur?

## Characteristics of earthquakes

The place where earthquakes begin, deep within the earth's crust, is called the **focus** (diagram **A**). Deep-focus earthquakes cause less damage and are felt less than shallow-focus ones. This is why the earthquake at Market Rasen in Lincolnshire on 27 February 2008, measuring 5.2 on the **Richter scale**, was felt so widely but not severely. The focus was 18.6 km below the surface. The point on the ground surface immediately above the focus is called the **epicentre**. Radiating out from this point are **shock waves**. These shock waves (seismic waves) cause the ground shaking that is responsible for much of the destruction caused by earthquakes.

## Measuring earthquakes

When earthquakes occur, seismographs record the extent of the shaking by a pen identifying the trace of the movement on a rotating drum. The line graph produced is called a seismogram (graph **B**).

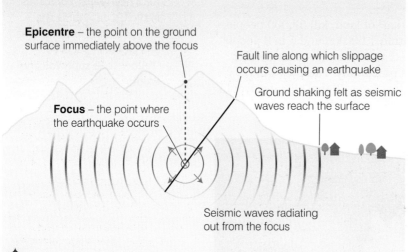

Epicentre – the point on the ground surface immediately above the focus

Fault line along which slippage occurs causing an earthquake

Ground shaking felt as seismic waves reach the surface

Focus – the point where the earthquake occurs

Seismic waves radiating out from the focus

**A**    *The features of an earthquake*

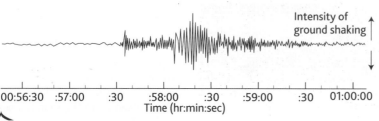

Intensity of ground shaking

00:56:30    :57:00    :30    :58:00    :30    :59:00    :30    01:00:00
Time (hr:min:sec)

**B**    *Seismogram of Market Rasen earthquake, 27 February 2008*

**In this section you will learn**

the features of an earthquake and how they are measured

where volcanoes occur and why they are found at constructive, destructive and conservative plate margins.

**Key terms**

**Focus**: the point in the earth's crust where the earthquake begins.

**Richter scale**: a scale ranging from 0 to 10 used for measuring earthquakes, based on scientific recordings of the amount of movement.

**Epicentre**: the point at the earth's surface directly above the focus.

**Shock waves**: seismic waves generated by an earthquake that pass through the earth's crust.

**Mercalli scale**: a means of measuring earthquakes by describing and comparing the damage done, on a scale of I to XII.

**Did you know** ??????

The most powerful earthquake ever recorded hit Valdivia in Chile in 1960 and measured 9.5 on the Richter scale. The two quakes that caused the most deaths occurred in China, with 830,000 people dying in Shensi in 1556 (over 8.0 on the Richter scale) and 255,000 (official figure) or 655,000 (unofficial figure) dying in Tangshan in 1976. The most powerful earthquake in the UK was 6.0 at Dogger Bank in 1931.

## The Richter scale

The strength of earthquakes is generally given according to the Richter scale. There is no upper limit to this scale. The logarithmic nature of the scale means that there is a 10-fold increase every time the scale increases by 1. So a scale 2 earthquake on the Richter scale is 10 times more powerful than a scale 1; a scale 3 earthquake is 10 times more powerful than a scale 2 and 100 times more powerful than a scale 1.

## The Mercalli scale

The **Mercalli scale** measures the effects of earthquakes using a scale from I to XII. It uses subjective descriptions of the resulting damage (table **C**).

## ■ Where and why do earthquakes occur?

Map **E** shows the location of areas prone to earthquakes. Look back at map **C** on page 7, and you will see a close link between plate margins and where earthquakes occur. The friction and pressures that build up where the plates meet are the causes of earthquakes.

- ■ **Destructive margins** – the pressure resulting from the sinking of the subducting plate and its subsequent melting can trigger strong earthquakes as this pressure is periodically released.

- ■ **Constructive margins** – here earthquakes tend to be less severe than those at destructive or conservative plate margins. The friction and pressure caused by the plates moving apart is less intense than at destructive plate margins.

- ■ **Conservative margins** – here, where the plates slide past each other, the plates tend to stick for periods of time. This causes stresses and pressure to build. The release of the pressure occurs in a sudden, quick release of the plates and often results in powerful earthquakes.

| **C** | The Mercalli scale |
|---|---|
| I | Barely felt |
| II | Felt by a few sensitive people; some suspended objects may swing |
| III | Slightly felt indoors as though a large truck were passing |
| IV | Felt indoors by many people; most suspended objects swing; windows and dishes rattle; standing cars rock |
| V | Felt by almost everyone; sleeping people are awakened; dishes and windows break |
| VI | Felt by everyone; some are frightened and run outside; some chimneys break; some furniture moves; slight damage |
| VII | Considerable damage in poorly built structures; felt by people driving; most are frightened and run outside |
| VIII | Slight damage to well-built structures; poorly built structures are heavily damaged; walls, chimneys and monuments fall |
| IX | Underground pipes break; foundations of buildings are damaged and buildings shift off foundations; considerable damage to well-built structures |
| X | Few structures survive; most foundations destroyed; water moved out of banks of rivers and lakes; avalanches and rockslides; railroads are bent |
| XI | Few structures remain standing; total panic; large cracks in the ground |
| XII | Total destruction; objects thrown into the air; the land appears to be liquid and is visibly rolling like waves |

**D**   *Buildings damaged by the Christchurch earthquake. What is the level of damage according to the Mercalli scale (Table **C**)?*

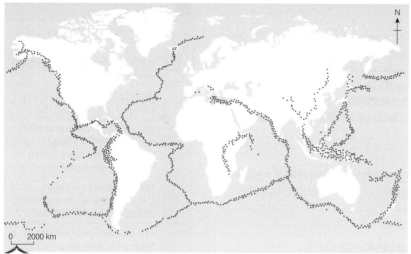

**Study tip**

Ensure that you can describe and explain the distribution of earthquakes – where they occur and why.

**⚭links**

There are many facts and figures available at **http://earthquake.usgs.gov** and **www.earthquakes.bgs.ac.uk**.

**E** *The location of earthquake zones*

## Activities

**1** Study diagram **A**.

  a  Draw your own diagram to illustrate:

   ■ the focus

   ■ the epicentre

   ■ the spread of the seismic (shock) waves.

  b  Include a definition of the terms next to your diagram.

**2** a  Make a bulleted list of the main points about the Richter scale.

  b  Illustrate the logarithmic scale by drawing a bar graph for 1 to 4 on a simple diagram of the scale.

  c  How many times more powerful is an earthquake measuring 7 than an earthquake measuring 5 on the Richter scale?

  d  How many times more powerful is an earthquake measuring 9 than an earthquake measuring 5 on the Richter scale?

**3** Describe the seismogram for the Market Rasen earthquake (graph **B**).

**4** Study table **C**.

  a  Select two scores on the Mercalli scale that are at least four apart and draw simple diagrams to illustrate the damage.

  b  What are the advantages and the disadvantages of the two methods of measuring earthquakes?

**5** Study map **E**, above, and map **C** on page 7.

  a  On an outline map of the world, shade in the areas that experience earthquakes.

  b  Label your map to describe the distribution of earthquakes. Use an atlas to help you with place names.

  c  Add the locations of the following recent earthquakes to your map:

   ■ Christchurch, New Zealand (2011)

   ■ Port-au-Prince, Haiti (2010)

   ■ Sichuan, China (2008)

   ■ Kashmir, Pakistan (2005)

   ■ Bam, Iran (2003)

   ■ Kobe, Japan (1995).

   Add any additional recent earthquakes.

  d  Provide evidence to support the hypothesis that earthquakes mostly occur at plate boundaries.

**6** Study diagrams **D**, **E** and **F** on pages 8 and 9, and the text on these pages.

  a  Draw simple, cross-sectional diagrams of each type of plate margin.

  b  Label your diagrams to explain where and why earthquakes occur at each plate margin.

  c  Give an example of an earthquake that has happened at each plate margin and name the plates responsible for it.

# 1.8   How do the effects of earthquakes differ in countries at different stages of development?

## The Kobe earthquake, Japan

At 5.46am on 17 January 1995, the Philippines Plate shifted uneasily beneath the Eurasian Plate along the Nojima fault line that runs beneath Kobe (diagram **A**). This collision of plates led to an earthquake measuring 7.2 on the Richter scale, with tremors lasting 20 seconds.

### In this section you will learn

case studies of earthquakes in countries at different stages of development – their causes, effects and responses

how and why the effects and responses are different in these two areas

the ways of trying to reduce the impact of earthquakes: the three Ps.

**1** Philippines Plate moves towards Eurasian Plate.

**2** Philippines Plate is forced down as it is oceanic crust.

**3** Plates jam together and pressure builds up.

**4** Pressure is suddenly released and plate jerks forward.

**5** Earthquake shockwaves travel outwards.

Akashi Bridge

Kobe

Osaka

To Tokyo

Rokko Island

Port Island

Osaka Bay

Kansai Airport

Awaji Island

Inland Sea

Eurasian Plate

Friction

Philippines Plate

Key

● Focus

◉ Epicentre

||| Shockwaves

☆ Major fire

**A**   *Timeline of the Kobe earthquake*

## The effects of the Kobe earthquake

In this short time, the earthquake claimed the lives of 6,434 people and seriously injured over 40,000. Some 300,000 were made homeless. Gas mains were ruptured, water pipes fractured, sections of elevated roads collapsed (photo **B**) and railway lines buckled. Two million homes were without electricity and one million people had to cope without water for 10 days.

Fires engulfed parts of the city, especially to the west of the port, devouring the wooden structures (photo **C**). Damage to roads and water supply made attempts to extinguish them impossible. People huddled in blankets on the streets and in tented shelters in parks in fear of returning to buildings damaged by the earthquake. The damage caused was in excess of $220 billion and the economy suffered. Companies such as Panasonic had to close temporarily.

**B**   *The Great Hanshin Expressway*

## Responses to the Kobe earthquake

Friends and neighbours searched through the rubble for survivors, joined by the emergency services when access was possible. Hospitals struggled to cope with the injured, treating people and operating in corridors. Major retailers such as 7-Eleven helped to provide essentials and Motorola maintained telephone connections free of charge. The railways were 80 per cent operational within a month. It took longer to restore the road network – most was operational by July, although it was not until September 1996 that the Hanshin Expressway was fully open again. A year later, the port was 80 per cent operational, but much of the container shipping business had been lost.

Buildings and structures that had survived the earthquake had been built to a 1981 code, whereas those that had complied with earlier 1960s practices had collapsed. This led to changes. New buildings were built further apart, to prevent the domino effect. High-rise buildings had to have flexible steel frames; others were built of concrete frames reinforced with steel instead of wood. Rubber blocks were put under bridges to absorb shocks.

**C**  *Fire in the area west of Kobe port*

**Did you know** ???????

The Japanese practise an earthquake drill every year to prepare them for an event such as that at Kobe. Over 800,000 people took part in a drill in August 2006.

## Case study

# The Haiti earthquake 2010

At 16:53 on 12 January 2010 the Caribbean island of Haiti was struck by a powerful 7.0 magnitude earthquake. The epicentre of the earthquake was just 15 km SW of the capital Port-au-Prince. The earthquake was caused by stress building up along the conservative plate margin marking the boundary between the North American plate and the Caribbean plate (map **D**). The stress was released by a sudden slippage along a fault (crack in the earth's crust) running parallel to the plate boundary just south of Port-au-Prince. Following the main earthquake there were several minor tremors measuring up to 5.0 on the Richter scale.

## The effects of the Haiti earthquake

The earthquake devastated large parts of the capital Port-au-Prince and resulted in a massive loss of life making it one of the most destructive earthquakes of all time.

- Approximately 230,000 people were killed.
- Over 2 million people were affected by the earthquake and 1.5 m were made homeless.
- About 180,000 homes were destroyed by ground shaking – this is an example of a primary effect (photo **E**).
- The homeless were accommodated in over 1,100 squalid camps with limited services such as water and sanitation. Many remained in these camps for well over a year (photo **F**).

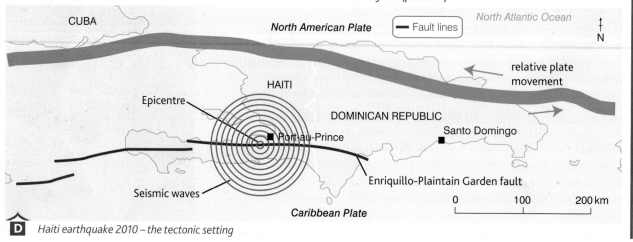

**D**  *Haiti earthquake 2010 – the tectonic setting*

- Cholera claimed the lives of several hundred people, mainly children in the aftermath of the earthquake. This is a good example of a secondary effect. Storms and flooding caused further hardship in the camps.
- The earthquake created some 19 million cubic metres of rubble and debris in Port-au-Prince – enough to fill a line of shipping containers stretching end to end from London to Beirut.
- Approximately 5,000 schools were damaged or destroyed.
- Services such as electricity, water, sanitation and communications were badly disrupted or destroyed.
- The total cost of repairing the damage done by the earthquake is $US11.5 bn over a period of 5–10 years.

**E**   *The impacts of the earthquake*

## Why did Haiti suffer so much?

Later in 2010, Christchurch in New Zealand was affected by an earthquake of exactly the same magnitude as the one that struck Haiti, 7.0 on the Richter scale. Yet, no one was killed. So why did Haiti suffer so much?

There are several reasons why Haiti suffered so much death and destruction:

- Haiti is an incredibly poor country. It was unprepared for an earthquake and could not cope adequately after the event. Before the earthquake most people survived on just $US2 a day.
- Over 80 per cent of the people in Port-au-Prince lived in poorly constructed, high density concrete buildings that simply fell apart when the earthquake struck.
- The earthquake was very close to the capital city Port-au-Prince and had a shallow focus. This meant that the city experienced very severe ground shaking.
- The port was largely destroyed and the airport badly damaged making it hard to bring in emergency supplies.

- The lack of a stable government resulted in limited and, at times, chaotic search and rescue efforts and recovery was very slow.
- There was a lack of doctors, hospitals (most were destroyed) and medical supplies. Many people died from their injuries or from diseases.

## Responses to the Haiti earthquake

The main **short-term response** involved search and rescue. Teams of specially trained medics with sniffer dogs and high tech heat-sensitive equipment were flown into the country to assist local people in rescuing those trapped by the collapsed buildings. Aid in the form of food, water, medical supplies and temporary shelter was brought in to the country from the USA and neighbouring Dominican Republic. The United Nations and USA provided security to maintain law and order and ensure a fair distribution of aid. The UK's Disasters Emergency Committee (DEC) raised over £100m. This money was used to support over 1.2 million people by providing emergency shelter, medical consultations, clean drinking water and sanitation.

In the weeks and months that followed there were several longer-term responses:

- Three-quarters of the damaged buildings were inspected and repaired.
- Some 200,000 people have received cash or food for public work, such as clearing the many tonnes of rubble.
- Several thousand people have decided to move away from Port-au-Prince to stay with family; some have even emigrated to other countries.
- The World Bank pledged $US100m to support reconstruction and recovery programmes in Haiti.

**F**   *Temporary camp set up to house the homeless in Port-au-Prince*

## Prediction, protection and preparation

**The three Ps** provide the key to trying to reduce the impact of earthquakes. **Prediction** involves trying to forecast when an earthquake will happen. Japan tries to monitor earth tremors with a belief that warning can be given, but this did not happen at Kobe. Foreshocks do occur, but not on a timescale useful to evacuation. Experts know *where* earthquakes are likely to happen, but struggle to establish *when*. Even looking at the time between earthquakes in a particular area does not seem to work. Similarly, experts struggle to pinpoint exactly where along a plate margin they will occur. Animal behaviour has been used in the east, but it is viewed sceptically in the USA. China evacuated the city of Haicheng (population 1 million) in 1975, partly due to the strange, unexplained behaviour of animals. Days later an earthquake struck, measuring 7.3 on the Richter scale. There were relatively few deaths, but it is estimated that 150,000 would have died without the evacuation.

Building to an appropriate standard and using designs to withstand movement is the main way of ensuring **protection** (photo **G** and diagram **H**). **Preparation** involves hospitals, emergency services and inhabitants practising for major disasters, including having drills in public buildings and a code of practice so that people know what to do to reduce the impact and increase their chance of survival.

**G** *The Transamerica Pyramid, San Francisco*

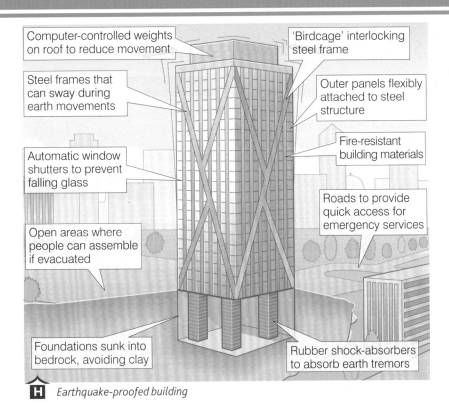

Computer-controlled weights on roof to reduce movement

'Birdcage' interlocking steel frame

Steel frames that can sway during earth movements

Outer panels flexibly attached to steel structure

Fire-resistant building materials

Automatic window shutters to prevent falling glass

Roads to provide quick access for emergency services

Open areas where people can assemble if evacuated

Foundations sunk into bedrock, avoiding clay

Rubber shock-absorbers to absorb earth tremors

**H**   *Earthquake-proofed building*

## ⬭links

To research Kobe, go to **www.seismo.unr.edu**; for Haiti go to **http://earthquake.usgs.gov**.

For information on earthquake preparation and drills, visit **www.bbc.co.uk** and **www.sfgate. com**. Enter 'earthquake drill' into the search box.

## Activities

**1**   Study diagram **A** and photos **B** and **C**.

a   Produce a fact file to summarise the main points about the Kobe earthquake such as the date, time, focus and epicentre.

b   Imagine you are the science editor on a newspaper. Explain the specific causes of the Kobe earthquake.

c   Imagine you are a resident of Kobe. Describe the effects of the earthquake, distinguishing between primary and secondary effects in your account. Make sure what you write relates specifically to Kobe.

d   Summarise the immediate and long-term responses to the earthquake in Kobe. Do this as a piece of writing, a table or a list of bullet points.

**2**   Copy the table below to summarise key facts about the two earthquakes.

| Feature | Kobe earthquake | Haiti earthquake |
|---|---|---|
| Cause | | |
| Primary effects | | |
| Secondary effects | | |
| Immediate responses | | |
| Long-term responses | | |

**3**   Describe how and explain why the effects of an earthquake differ in countries at different stages of development.

**4**   For each of prediction, protection and preparation, answer the following questions.

a   Explain the meaning of the term.

b   Give an example of the method for reducing the impact of earthquakes from either Kobe or Haiti.

c   Give one advantage and one disadvantage of the method.

d   Which of the three Ps do you think is the most useful in reducing the impact of earthquakes? Justify your choice.

**5**   Draw an illustrated earthquake code for a school in an earthquake-prone area.

## 1.9    Why is a tsunami hazardous?

### How tsunamis form

**Tsunamis** are usually triggered by earthquakes. The crust shifting is the primary effect; a knock-on (secondary) effect of this is the displacement of water above the moving crust. This is the start of a tsunami.

A normal, wind-driven wave may have a length of 100 m from crest to crest, but a tsunami may be 200 km in length. The heights also greatly differ: 2 m for a normal wave versus 1 m for a tsunami out at sea. Tsunamis move at speeds of around 800 kph, rapidly approaching the coast almost unnoticed. As they near land they slow, reduce in length and gain in height (diagram **A**).

> **In this section you will learn**
>
> what a tsunami is and why it is a secondary effect of an earthquake
>
> the causes and effects of, and responses to, a tsunami.

> **Key term**
>
> **Tsunami:** a special type of wave where an event, often an earthquake, moves the entire depth of the water above it.

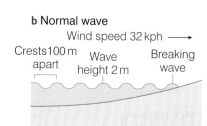

**a Tsunami**

Crest only 1 m high

Crests 200 km apart

Height increases

Wavelength shortens

Water travelling at 800 kph

30 m

Slip on fault line causes water above to move upwards

**b Normal wave**

Wind speed 32 kph →

Crests 100 m apart

Wave height 2 m

Breaking wave

**A**   *Comparing wind-driven waves and tsunamis*

---

**Case study**

## The Tohoku tsunami, Japan 2011

In the space of just a few years powerful tsunami have caused widespread destruction and huge loss of life. On 26 December 2004 a massive earthquake in the Indian Ocean measuring 9.3 on the Richter scale triggered a tsunami up to 25 m high that swept onshore killing an estimated 220,000 people in Indonesia, India and Sri Lanka.

On 11 March 2011, another very powerful earthquake measuring 9.0 on the Richter scale occurred about 100 km east of Sendai on Honshu, Japan. In just 30 minutes a wall of water up to 40 m high hit the coast of north western Japan. It was followed in places by up to 9 additional 'waves' of up to 10 m in height. Some 3,000 km of coastline were affected by the waves, which tore through coastal defences and inundated the coastal plains.

**B**   *Aerial view of the impact of the Japanese tsunami*

### Causes

The earthquake that triggered the tsunami occurred at the destructive plate margin where the Pacific plate is being subducted beneath the North American plate. Scientists believe that a segment of rock, some 200 km in length slipped suddenly resulting in an upwards 'flick' of the earth's crust by between 5–10 m. It was this sudden uplift that triggered the tsunami.

### Effects

There were a number of effects of the tsunami:

- Over 20,000 people were killed as the waves swept up to 10 km onshore. The high death toll was due to the power of the surge of water which overtopped tsunami defences and flooded areas thought to be safe from tsunami.

- Some 500 km² of coastal plains were inundated, destroying farmland, settlements and communications. The port city of Sendai (population 100,000) was virtually destroyed. A total of 200,000 buildings were damaged or destroyed by the earthquake and tsunami.

- Ruptured gas pipes led to fires that raged for several days.

- Explosions occurred at the Fukushima nuclear power plant as seawater over-topped the flood defences. There were considerable concerns about nuclear contamination and the possibility of a meltdown as the cooling systems failed to operate.

- Electricity was cut off in almost six million homes and over one million people were left without running water.

- Heavy snow, roads blocked by debris and landslides and over 1,000 aftershocks hampered relief efforts. In some areas, stocks of food, water and medical supplies ran low.

- Stock markets around the world fell over concerns about Japan's rising debts in the face of billions of US$ worth of damage.

## Responses

- Over 100,000 Japanese soldiers were deployed in search and rescue. They distributed blankets, water and food to the people affected by the disaster.

- Specialist search and rescue teams were flown in to the area from overseas.

- An exclusion zone was set up around the Fukushima nuclear plant and people were evacuated from the area.

- In the longer-term, a huge re-building and reconstruction programme is planned involving houses, infrastructure and communications systems (roads, railways, etc.). Port facilities will need to be re-built. The system of tsunami defences will need to be reconsidered and may well be extended in height beyond the standard 12 m that is currently the accepted level.

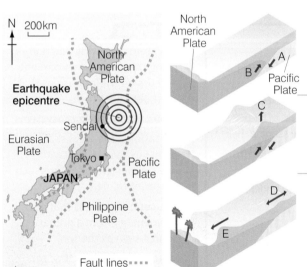

**C**  *The cause of the Tohoku tsunami*

1 One tectonic plate is dragged beneath the other causing the edges of the plates to flex and deform causing pressure to build over time.
  A Plate subducts beneath the other
  B Pressure builds up gradually causing distortion

2 The massive stresses eventually cause the plate to snap back displacing a vertical column of water which causes a tsunami.
  C Fracture causes water displacement

3 At the fault line the tsunami is merely a ripple but as it approaches land it gains height as the water becomes shallower slowing the wave.
  D Column of water splits into two; one travels out to sea, the other towards land
  E Wave height increases as it reaches shallower water

## Activity

Working in small groups of three or four, produce a report or presentation on the Japanese tsunami to include the following:

a A definition of 'tsunami', supported by a labelled diagram.

b The cause of the tsunami from a science editor's viewpoint, supported by a labelled diagram.

c The effects of the tsunami from a resident's or tourist's point of view. Include a labelled photograph showing the impact and a map showing the areas affected.

d The immediate responses to the tsunami – a resident's, doctor's or Red Cross volunteer's eyewitness account. Try to capture the feelings of the person involved.

e The long-term responses to the tsunami – the views of a resident as rebuilding continues, or of a government or aid agency worker.

f A summary of 10 important facts about the tsunami.

It must be clear that you are writing about the Japanese tsunami. You should try to convey the scale of the disaster and the human suffering.

# 2 Ecosystems and global environments

## 2.1 What is an ecosystem?

An **ecosystem** is a natural system that comprises plants (flora) and animals (fauna) and the natural environment in which they live. There are often complex relationships between the living and non-living components in an ecosystem. Non-living components include the climate (primarily temperature and rainfall), soil, water and light.

Ecosystems can be identified at different scales. A local ecosystem can be a pond (diagram **A**) or a hedge. Larger ecosystems can be lakes or woodlands. It is possible to identify ecosystems on a global scale, such as tropical rainforests or deciduous woodland. These global ecosystems are called **biomes**.

### In this section you will learn

the concept of an ecosystem and its key components, such as producers, consumers and food chains

how change can have a considerable effect in an ecosystem.

## Case study

### The freshwater pond ecosystem

Freshwater ponds provide a variety of habitats for plants and animals (diagram **A**). Note that there are considerable variations in the amount of light, water and oxygen available in different parts of a pond. Animals living at the bottom in deep water need different **adaptations** to those living on the margins of the pond. Certain plants such as water lilies tolerate total immersion by sending their flowering stems to the surface of the water. Reeds and other similar plants are better adapted to being right on the edges as they can tolerate drier conditions.

There are a number of important ecological concepts that you need to understand:

- **Producers** and **consumers** – organisms can be either producers or consumers. Producers convert energy from the environment (typically sunlight) into sugars (glucose). The most obvious producers are plants, which convert energy from the sun by the process of photosynthesis. Consumers obtain their energy from the sugars made by the producers. The grasses at the margins of the pond in diagram **A** are good examples of producers. A pond snail is a good example of a consumer because it eats the plants.

- **Food chain** – this shows the links (hence the term 'chain') between producers and consumers. Diagram **B** shows a food chain that might exist in a typical pond. Note that it is a simple linear series of connections.

- **Food web** – this shows the connections between producers and consumers in a rather more detailed way, hence the term 'web' rather than 'chain' (diagram **C**).

- **Scavengers** and **decomposers** – when living elements (plants and animals) of an ecosystem die, scavengers and decomposers break them down and effectively recycle their nutrients. Scavengers eat dead animals and plants. A rat-tailed maggot is a good example of a freshwater pond scavenger. Flies and earthworms are examples of scavengers found on land. Decomposers are usually bacteria and fungi. They break down the remaining plant and animal material, often returning the nutrients to the soil.

### Key terms

**Ecosystem:** the living and non-living parts of an environment and the interrelationships that exist between them.

**Biomes:** global-scale ecosystems.

**Adaptations:** the ways that plants evolve to cope with environmental conditions such as lots of rainfall.

**Producers:** organisms that get their energy from a primary source such as the sun.

**Consumer:** organisms that get their energy by eating other organisms.

**Food chain:** a line of linkages between producers and consumers.

**Food web:** a diagram that shows all the linkages between producers and consumers in an ecosystem.

**Scavengers:** organisms that consume dead animals or plants.

**Decomposers:** organisms such as bacteria that break down plant and animal material.

**Nutrient cycling:** the recycling of nutrients between living organisms and the environment.

■ **Nutrient cycle** – nutrients are foods that are used by plants or animals to grow, such as nitrogen, potash and potassium. There are two main sources of nutrients: rainwater washes chemicals out of the atmosphere and weathered rock releases nutrients into the soil. When plants or animals die, the scavengers and decomposers recycle the nutrients, making them available once again for the growth of plants or animals.

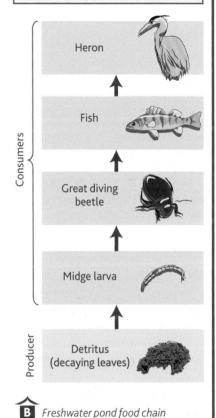

**Did you know** ???????

The rat-tailed maggot has a telescopic breathing siphon in its posterior. As an adult it becomes a drone fly, which looks similar to a honey bee.

**B**  *Freshwater pond food chain*

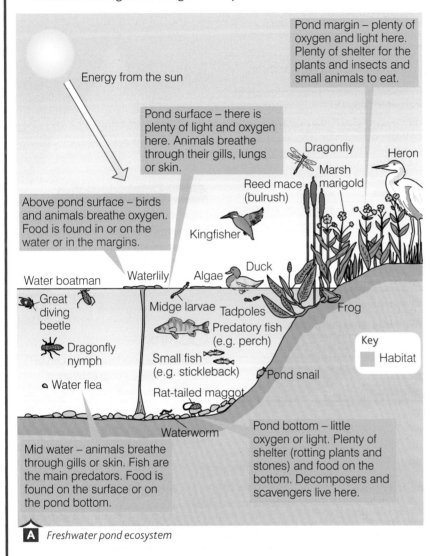

**A**  *Freshwater pond ecosystem*

## The impact of change on the freshwater pond ecosystem

The diversity and relative numbers of the components in an ecosystem can change over time. This can be caused by natural factors such as environmental change (e.g. flood, fire, drought) or human-induced change (e.g. drainage, reclamation, fish stocking). Once a change has occurred it is rarely isolated and often has an impact on other parts of the ecosystem.

If predatory fish are introduced into the pond in diagram **A** they will eat more of the smaller fish and small animals such as frogs. This will affect the numbers of those creatures, which will in turn reduce the amount of food available to creatures further up the food chain. At the same time, with fewer frogs in the pond, numbers of those creatures below frogs in the food chain such as slugs will increase.

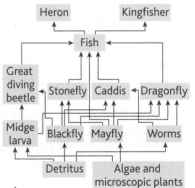

**C**  *Freshwater pond food web*

**D**  *Freshwater pond species and energy sources*

| Species | Energy source (sunlight or food) |
| --- | --- |
| Algae | Sunlight |
| Dragonfly | Other adult insects |
| Dragonfly nymph | Tadpoles, young fish, water fleas, beetles |
| Duck | Water plants, insects, tadpoles, small fish, pond snails |
| Frog | Insects, water worms, snails |
| Great diving beetle | Water fleas, midge larvae, pond snails, nymphs, tadpoles, water boatmen |
| Heron | Fish, frogs and tadpoles, larger insects |
| Kingfisher | Small fish, tadpoles, small frogs, great diving beetle |
| Marsh marigold | Sunlight |
| Midge larvae | Microscopic plants, small particles of dead plants |
| Perch | Small fish, beetles, water fleas |
| Pond snail | Large water plants, algae |
| Rat-tailed maggot | Decaying plants |
| Reed mace | Sunlight |
| Sticklebacks | Tadpoles, young fish, water fleas, beetles |
| Tadpole | Microscopic plants, algae, midge larvae |
| Water boatmen | Tadpoles, water worms, midge larvae, water fleas |
| Water flea | Microscopic plants, small particles of dead plants |
| Water lily | Sunshine |
| Water worm | Small particles of dead animals |

**Study tip**

It is important that you learn the terminology of ecosystems and understand the principles of change in an ecosystem.

**⚭ links**

Excellent ecosystems links can be found at **www.geography.pwp. blueyonder.co.uk**.

**Activities**

**1**  Study diagram **A** and table **D**.

a  Identify some producers that live in a freshwater pond.

b  From where do these producers obtain their energy?

c  Bacteria and fungi are decomposers in a pond ecosystem. Can you name some scavengers?

d  Which pond species are at the top level in the food chain, i.e. they are not eaten by any other species in the ecosystem?

e  Select one of the species listed in **D** and draw a food chain diagram with your chosen species at the top. Include sunlight in your diagram and add sketches to make it look more interesting.

**2**  Study diagram **A** and table **D**.

a  In pairs, draw a food web for the species shown in diagram **A** and listed in table **D**.

b  Identify the producers in your food web.

c  Add some simple sketches or photos using the internet if you wish.

**3**  Study diagram **A** and table **D**.

a  Imagine that the landowner cuts down all the vegetation at the side of the pond to create a wooden deck for fishing. How would this affect the ecosystem in the short term and the long term?

b  Imagine that disease wipes out all the frogs. How would this affect the ecosystem in the short term and the long term?

c  Suggest another change that could happen to this ecosystem and describe the effects that this change might have on the species living in the pond.

# 2.2 What are the characteristics of global ecosystems?

## The distribution of global ecosystems

Global ecosystems are known as biomes. The dominant type of vegetation cover usually defines a biome. Map **A** shows the global distribution of the major world biomes.

The biome in the UK is **temperate deciduous forest**. This is the natural vegetation that would occur in much of the UK in response to climates and soils. It does not mean that the entire country is covered by woodland. However, if no land management took place at all in the UK for 100 years or so, then the landscape would start to revert back to natural deciduous woodland.

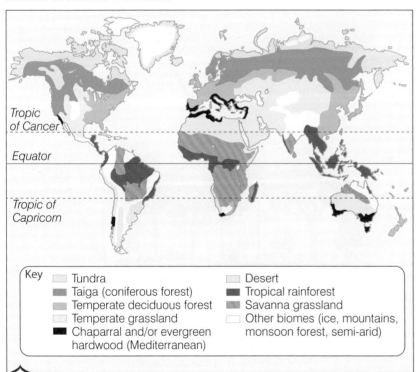

Key
- ☐ Tundra
- ■ Taiga (coniferous forest)
- ☐ Temperate deciduous forest
- ☐ Temperate grassland
- ■ Chaparral and/or evergreen hardwood (Mediterranean)
- ☐ Desert
- ■ Tropical rainforest
- ■ Savanna grassland
- ☐ Other biomes (ice, mountains, monsoon forest, semi-arid)

**A** Global biomes

**Key term**

**Temperate deciduous forest:** forests made up of broad-leaved trees such as oak that drop their leaves in the autumn.

## Temperate deciduous forests

Temperate deciduous forests are found across much of north-west Europe, eastern North America and parts of East Asia. They occur in these regions because they are well suited to the moderate climate (graph **B**). Rainfall is distributed evenly throughout the year, summers are warm but not too dry, winters are cool but not too cold. There is a long growing season lasting up to seven months.

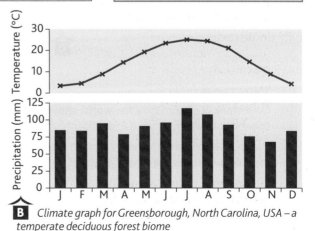

**B** Climate graph for Greensborough, North Carolina, USA – a temperate deciduous forest biome

Soils that develop under these climatic conditions tend to be rich and fertile. Weathering is active, providing plenty of nutrients, and the annual leaf fall provides organic matter to enrich the soil further. The common soil found in this biome is a brown soil.

Look at photo **C**, which shows oak woodland in southern England. This is typical temperate deciduous woodland. The main feature of the trees in this biome (e.g. oak, beech, birch, ash) is that they shed their leaves in autumn. This is what the term 'deciduous' means. They drop their leaves in response to reductions in light and heat, which enables them to conserve water. Deciduous trees are typically broad-leaved, which means there is a great deal of potential for water loss through the holes (stomata) on the underside of their leaves. Leaf fall comes early in some years if there has been a shortage of water in late spring and summer.

**C**   *A typical oak woodland in southern England*

Deciduous woodlands are rich in their diversity of vegetation and they provide a great range of habitats for the many plants and animals that live there (diagram **D**). One typical characteristic of deciduous woodland is the layering or **stratification** of the vegetation:

- The top of the fully grown trees provides a canopy, which acts like an umbrella. The main trees forming this layer are oak and ash.
- Beneath this is a sub-canopy of saplings and smaller trees, such as hazel.
- Below this is a herb layer of brambles, bracken, bluebells, wild garlic and ivy.
- Finally there is a ground layer close to the soil surface. Here, it is damp and dark – ideal conditions for moss to grow.

For much of the year it is quite dark in deciduous woodland, which is not ideal for flowering plants. This helps to explain why bluebells, for example, commonly flower in the early spring before the canopy has fully developed.

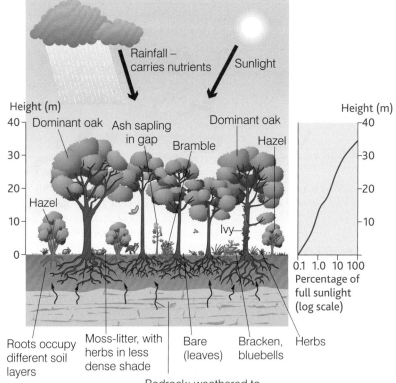

**D**   *A temperate deciduous forest ecosystem*

**Activities**

**1**   Study map **A**.

a   On a blank outline map of the world, show the distribution of temperate deciduous forest.

b   Use an atlas to identify some of the main regions and countries where this is the natural type of vegetation.

c   Why is a map of natural vegetation zones (biomes) slightly misleading? Consider the situation of the UK.

**2**   Study graph **B**.

a   Describe the climate of Greensborough, North Carolina.

b   Why is temperate deciduous forest well suited to this climate?

c   One of the main characteristics of deciduous trees is that they shed their leaves. Explain why this occurs and why it benefits the trees.

## Activity

**3** Study diagram **D**.

a What is the evidence from the diagram that the temperate deciduous forest supports a great diversity of wildlife?

b What is meant by stratification and how is this exhibited in a deciduous forest? Draw a diagram to support your answer.

c Why do you think stratification exists?

d How have flowers like bluebells adapted to living in a deciduous forest?

## Did you know ??????

The tallest tree in the world is a coast redwood found in Redwood National Park, California, which measures an extraordinary 115.55 m. In 2008 scientists claimed to have discovered the oldest tree in the world in Sweden, a Norway spruce said to be 9,550 years old.

## ■ Tropical rainforests

**Tropical rainforests** are found in a broad belt through the tropics (map **A**), from Central and South America, through central parts of Africa, in South-east Asia and into the northern part of Australia. This biome is characterised by a plentiful supply of rainfall (over 2,000 mm a year) and high temperatures (averaging 27°C) throughout the year. This climate (graph **E**) provides ideal conditions for plant growth.

Tropical rainforests have extremely lush and dense vegetation. If you were to enter a rainforest, you would need a torch and good shoes as it is dark and damp. The trees in a tropical rainforest grow to be extremely tall, often up to 45 m in height. There is a great variety of species, typically up to 100 in a single hectare. This explains why the wood is such a valuable resource.

As with deciduous woodlands, a tropical rainforest has a clear stratification (diagram **F**). It is interesting to note that, unlike deciduous forests where most plants and animals live close to the forest floor, in a tropical rainforest the majority are found in the canopy where there is maximum light. Some tree leaves are specially adapted to twist and turn to face the sun as it arcs across the sky. In contrast, rainforest floors are often too dark to support many plants.

## Key terms

**Stratification:** layering of forests, seen particularly in temperate deciduous forests and tropical rainforests.

**Tropical rainforests:** the natural vegetation found in the tropics, well suited to the high temperatures and heavy rainfall of these latitudes.

**Leaching:** the dissolving and removal of nutrients from the soil, often in tropical rainforests because of the heavy rainfall.

Tropical rainforests support the largest number of plant and animal species of any biome. The constant environmental conditions of its climate promote plant growth and result in a great variety of food sources and natural habitats. Many birds live in the canopy. Some mammals, such as monkeys, are well adapted to living in the trees. Animals such as deer live on the forest floor eating seeds and berries.

Tropical rainforest soils are surprisingly infertile considering the lush growth of the vegetation. Most of the nutrients are found at the surface where dead leaves decompose rapidly in the hot and humid conditions. Many of the trees and plants have shallow roots to absorb these nutrients and fungi growing on the roots transfer nutrients straight from the air. The heavy rainfall quickly dissolves and carries away nutrients. This is called **leaching**. It leaves behind an infertile red-coloured soil called latosol, which is rich in iron (hence the colour) and very acidic (photo **G**).

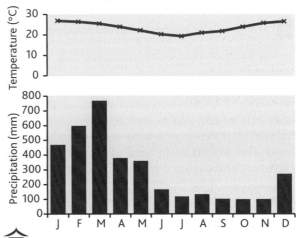

**E** *Climate graph for Innisfail, Queensland, Australia – a tropical rainforest*

Fast-growing trees such as capoc out-compete other trees to reach sunlight – such trees are called emergents

Many leaves have flexible bases so that they can turn to face the sun

Many leaves have a 'drip tip' to allow the heavy rain to drip off the leaf

Water drips off leaves

**G**    *Latosol: a typical soil found in the tropical rainforest biome*

Shrub layer and ground layer

Thin, smooth bark on trees to allow water to flow down easily

Buttresses – massive ridges help support the base of the tall trees and help transport water. May also help oxygen/carbon dioxide exchange by increasing the surface area

Lianas – woody creepers rooted to the ground but carried by trees into the canopy where they have their leaves and flowers

Plants called epiphytes can live on branches high in the canopy to seek sunlight – they obtain nutrients from water and air rather than soil

    *Stratification and vegetation adaptations in a tropical rainforest*

## ⚭ links

For more information on plant adaptations in arid environments, go to **www.cwnp.org/adaptations. html**.

## ▮ Hot deserts

A desert is an area that receives less than 250 mm of rainfall per year. The resulting dryness or **aridity** is the main factor controlling life in the desert. **Hot deserts** are generally found in dry continental interiors in a belt at approximately 30°N and 30°S. It is at these latitudes where air that has risen at the Equator descends, forming a persistent belt of high pressure (anticyclone). This explains the lack of cloud and rain and high daytime temperatures. It also explains why, with the lack of cloud cover, temperatures can plummet to below freezing at night during the winter.

Desert soils tend to be sandy or stony, with little organic matter due to the general lack of dense vegetation. Soils are dry but can soak up water rapidly after rainfall. Evaporation draws salts to the surface, often leaving a white residue on the ground. Desert soils are not particularly fertile.

## Activities

**4** Study graph **E**.

a   How does the climate of a tropical rainforest compare with that of a temperate deciduous forest?

b   How is the climate ideal for the growth of plants?

c   How does the climate provide opportunities as well as problems for the plants and animals living in this environment?

**5** Study diagram **F**.

a   What name is given to the tall trees that break through the canopy?

b   How high can the tallest trees grow?

c   On the ground in a tropical rainforest, in which layer of the forest would you be in?

d   How have the leaves of the tallest trees adapted to gain maximum sunlight?

e   Describe how leaves are designed to shed water quickly during torrential downpours.

f   What are lianas and how have they adapted to live successfully in tropical rainforests?

g   What is a buttress and what are the possible reasons why some trees have them?

h   Select and write about one other plant adaptation in a tropical rainforest.

**6** Study photo **G**.

a   What name is given to a tropical rainforest soil?

b   Describe the characteristics of the soil in the photo.

c   Why are most of the nutrients found near the surface of the soil?

d   How have plants adapted to this?

e   What is leaching and why is it a problem?

**7** In the past, some people cut down rainforest trees and replaced them with commercial crops expecting a wonderful harvest. Instead, the new plants grew poorly. From what you have learned, explain why this happened.

**8** Use the information in fact file **H**, together with your own internet research and the text here, to complete a short research project on desert ecosystems. You should include the following information:

■   Draw a map to show the main areas of hot desert. Use an atlas to name the deserts.

■   Using a climate graph, describe the climatic conditions experienced in hot deserts. Why are hot deserts hostile environments?

■   With the aid of labelled photos or sketches, describe the adaptations of plants and animals to hot desert conditions. A good website to get you started is www.cwnp.org/adaptations.html.

## Death Valley, California

|  | J | F | M | A | M | J | J | A | S | O | N | D |
|---|---|---|---|---|---|---|---|---|---|---|---|---|
| Average temperature (°C) | 11 | 15 | 19 | 24 | 29 | 35 | 38 | 37 | 32 | 25 | 17 | 10 |
| Average precipitation (mm) | 0.8 | 1.2 | 0.8 | 0.4 | 0.2 | 0.1 | 0.3 | 0.2 | 0.4 | 0.3 | 0.6 | 0.4 |

## Plant adaptations

1   Desert yellow daisy – small linear leaves that are hairy and slightly succulent.

2   Great basin sagebrush – tap roots up to 25 m long and small needle-like leaves to reduce water loss.

3   Giant saguaro cactus – roots very close to the surface so that it can soak up water before it evaporates. Outside skin is pleated so that it can expand when water is soaked up. Grows very slowly.

4   Joshua tree – needle-like leaves coated with a waxy resin.

 1
 2
 3
 4

**H**  *Desert ecosystem fact file*

# 2.3  What are temperate deciduous woodlands used for?

## Epping Forest, Essex

Epping Forest is an ancient deciduous forest that runs north-east of London on a high gravel ridge. It covers an area of about 2,500 ha and is about 19 km long and 4 km wide. It is the largest area of public open space near London.

Although 70 per cent of Epping Forest is deciduous woodland (mostly beech), there are a number of other natural environments including grasslands and marshes. It is home to a rich variety of wildlife including all three native species of woodpeckers and wood-boring stag beetles. Fallow deer still roam the forest.

### Early uses and management

Since Norman times, kings and queens of England have used Epping Forest for hunting deer. Local people ('commoners') were able to use the forest to graze their animals and to collect wood for firewood and building.

For many years the practice of **pollarding** was used to manage the woodland. This involves cutting the trees at about shoulder height, above the level of browsing by animals such as deer (photo **A**). Pollarded trees reshoot at this height, thereby producing new wood for future cutting. This is a good example of **sustainable management** as it ensures a supply of wood for future generations. It also accounts for the presence of some ancient trees because, rather than being felled for timber, they were pollarded.

As royal use declined in the 19th century, local landowners made attempts to buy parts of the forest. In response to this threat, in 1878 the Epping Forest Act of Parliament was passed in which it was stated that 'the Conservators shall at all times keep Epping Forest unenclosed and unbuilt on as an open space for the recreation and enjoyment of the people'.

Since 1878 the Forest has been managed by the City of London Corporation.

### Recent management

Epping Forest is an excellent example of a natural deciduous forest that is being managed sustainably for the future. Over 1,600 ha of the forest has been designated a Site of Special Scientific Interest and a European Special Area of Conservation. This offers protection under law to its large number of ancient trees, which support a vast variety of flora and fauna.

**A**  *An ancient pollarded tree, Epping Forest*

The overall planning responsibility of Epping Forest lies with the City of London Corporation, which produces management plans to ensure that the forest continues to provide open space for the public while conserving the natural environment. Planning measures adopted include the following:

- managing recreation by providing appropriate car parks, toilets and refreshment facilities and by maintaining footpaths (photo **B**)

### In this section you will learn

the various uses of a temperate deciduous forest

how deciduous forests can be managed sustainably.

### Key terms

**Pollarding:** cutting off trees at about shoulder height to encourage new growth.

**Sustainable management:** a form of management that ensures that developments are long lasting and non-harmful to the environment.

### ⚲links

Further information about Epping Forest can be found at **www.bbc.co.uk**. Type 'Epping forest' into the search box.

The Epping Forest Information Centre can be found at High Beach, Loughton, Essex IG10 4AF.

### Did you know ??????

Epping Forest is a renowned location for the rare stag beetle due to the presence of dead and decaying wood. The stag beetle is one of several species of wildlife that are considered to be rare or endangered.

### Study tip

Take time to learn the various measures of sustainable management that have been adopted, both past and present.

- providing three easy-access parks to allow access for people with disabilities
- allowing old trees to die and collapse naturally unless they are dangerous
- controlling some forms of recreation, such as riding and mountain biking, which may damage or affect other forms of recreation
- preserving ancient trees by re-pollarding them to enable new shoots to grow – since 1981, over 1,000 ancient trees have been re-pollarded
- encouraging grazing to maintain the grassland and the flora and fauna associated with it
- preserving ancient earthworks and buildings
- maintaining ponds to prevent them silting up
- preserving the herd of fallow deer.

**B**  *Recreation in Epping Forest*

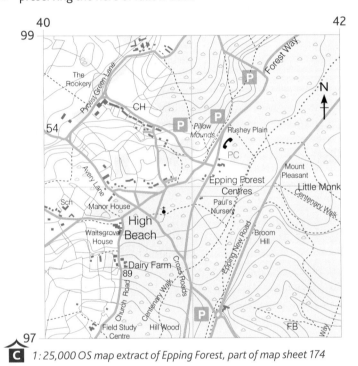

**C**  *1:25,000 OS map extract of Epping Forest, part of map sheet 174*

Key

°° Non-coniferous trees

P Car park

0 ——————— 500 m

## Activities

**1** Study photo **A**.

a What is meant by pollarding? Draw a sketch of the tree in the photo to illustrate your answer.

b Why are trees pollarded?

c Why is pollarding an example of sustainable management?

d How does pollarding lead to the survival of ancient trees?

e Suggest why trees are being re-pollarded today.

**2** Study map extract **C**.

a Identify the different types of natural environment.

b What are the attractions and opportunities for recreation?

c Suggest any conflicts that might arise between people visiting Epping Forest.

d Why do you think it is important to have properly designated car parking?

e You may have noticed that there is a field study centre. Why is this a good location for an education centre?

**3** Do you think the Forest is being managed sustainably? Explain your answer.

# 2.4    Deforestation in Malaysia

## Malaysia's tropical rainforests

Malaysia is a country in south-east Asia. It is made up of Peninsular Malaysia and Eastern Malaysia, which is part of the island of Borneo (maps in **A**). Along with neighbouring countries, the natural vegetation in Malaysia is tropical rainforest. Nearly 60 per cent of Malaysia is forested and commercial tree crops, primarily rubber and oil palm, occupy a further 13 per cent. Trees and forest cover an area equivalent to the whole of the UK.

**a** World location map

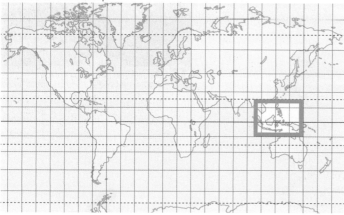

**b** Regional location map

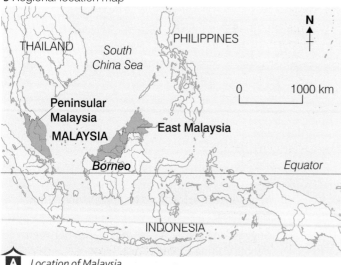

**A**  *Location of Malaysia*

In the past most of the country was covered by **primary (virgin) rainforest**. In Peninsular Malaysia most of this has now gone and little is left on Borneo. Today an estimated 18 per cent of Malaysia's forest is virgin forest.

Malaysian rainforests support over 5,500 species of flowering plants (the UK has 1,350), 2,600 species of tree (UK 35) and over 1,000 species of butterflies (UK 43). Of the 203 species of mammals, 78 per cent live only in forests. Malaysia's rainforests are clearly special places.

### In this section you will learn

the causes of deforestation in tropical rainforests

the effects of deforestation in tropical rainforests.

### Key terms

**Primary (virgin) rainforest:** rainforest that represents the natural vegetation in the region unaffected by the actions of people.

**Deforestation:** the cutting down and removal of forest.

**Clear felling:** absolute clearance of all trees from an area.

**Selective logging:** the cutting down of selected trees, leaving most of the trees intact.

### Did you know ??????

The rainforests of South-east Asia are the heartland for the giant dipterocarp trees, of which there are 515 species. These tall, straight trees dominate the tropical timber trade. They are only found in virgin forests and form vital vertical structures in the forests by supporting many species of wildlife.

## ◼ Threats to Malaysia's rainforests

Recent statistics from the United Nations (UN) suggest that the rate of **deforestation** in Malaysia is increasing faster than in any other tropical country in the world, increasing 85 per cent between the periods 1990 to 2000 and 2000 to 2005. Since 2000, some 140,200 ha of forest have been lost on average every year. There are several threats to the rainforests in Malaysia.

### Logging

During the 1980s, rampant logging on Borneo led to Malaysia becoming the world's largest exporter of tropical wood. **Clear felling**, where all trees are felled in an area, was common and this led to the total destruction of forest habitats. In recent years the main logging practice has been **selective logging**. Theoretically only fully grown trees are felled and those with important ecological qualities are left unharmed. Although selective logging is far less damaging, it reduces biodiversity. All forms of logging require road construction to bring in machinery and take away the timber (photo **B**).

**B**   *Road construction and logging in Sarawak, East Malaysia*

Malaysia has one of the best rainforest protection policies in the region, but environmental groups claim to have found evidence of illegal logging in Borneo. Here, increasingly marginal slopes have been logged, leading to problems of soil erosion and mudslides.

Not only has logging reduced biodiversity, it has also threatened indigenous tribes. In 2003 a local Penan community in the village of Long Lunyim in Sarawak state protested against the encroachment by a logging company. Some members of the community were imprisoned for their protests, and the company pushed on to the community's forest reserve to exploit the timber.

### Energy

The $2bn Bakun Dam project in Sarawak, due for completion in 2012, will result in the flooding of thousands of hectares of forest in order to supply hydroelectric power mainly for industrialised Peninsular Malaysia. An estimated 230 km$^2$ of virgin rainforest will have to be cut down for the project. Some 10,000 indigenous people have been forced to move from the flooded area. They are traditional subsistence farmers with little money, yet they are being asked to pay to be rehoused. Many now suffer from depression and alcoholism is rife.

### Mining

Mining has been widespread in Peninsular Malaysia, with tin mining and smelting dominating. Areas of rainforest have been cleared to make way for mining operations and the construction of roads. In some places, the mining activities have led to pollution of the land and rivers. Drilling for oil and gas has started on Borneo.

**Did you know** ??????

The orang-utan is a great ape found only in south-east Asia. Able to reason and think, it is one of our closest relatives, sharing 97 per cent of our DNA. It is the largest tree-living mammal in the world.

## Commercial plantations

Malaysia is a major producer of oil palm and rubber. In the early 20th century, forest was cleared to make way for the rubber plantations. In recent decades, however, synthetic rubber has led to a steep decline in rubber exports and many plantations have either been abandoned or converted to oil palm (photo **C**).

Today, Malaysia is the largest exporter of palm oil in the world. During the 1970s, large areas of land were converted to palm oil plantations. With plantation owners receiving a 10-year tax break, increasing amounts of land have been converted to plantations. Deforestation for palm oil is taking place on Borneo and threatening the survival of many species of wildlife including the orang-utan.

## Resettlement

In the past, poor urban dwellers were encouraged to move into the countryside to relieve pressure on cities. This policy is called **transmigration**. Between 1956 and the 1980s, an estimated 15,000 ha of rainforest was felled to accommodate the new settlers, many of whom set up plantations.

## Fires

Fires are common on Borneo. Some are natural, resulting from lightning strikes, whereas others result from forest clearance or arson. Occasionally, '**slash and burn**' agriculture – where local people clear small areas of land in order to grow food crops – results in wildfires.

**C**    *Oil palm plantation in Malaysia*

**Activities**

1  Study this section on the threats to Malaysia's rainforest.

a  Briefly outline the main causes of deforestation in Malaysia.

b  To what extent do you think deforestation in Malaysia has been driven by economic gain (i.e. making money)?

c  What have been the environmental effects of deforestation?

d  What have been the social effects (i.e. on the people) of deforestation?

e  Despite government policies to preserve Malaysia's rainforests, why do you think deforestation continues to be an issue in Malaysia?

2  Complete a large revision diagram to summarise the threats to Malaysia's rainforests, using an A3 sheet of paper if possible. Include some internet research information if you can.

a  At the centre of your diagram, place a photo or sketch to show the features of Malaysia's rainforest. You could include a map too.

b  Around the central feature, create a series of illustrated text boxes describing the main threats to the rainforest. Use arrows to link these boxes to the central image. Use plenty of colour and think carefully about your design to ensure that the final outcome supports your revision.

**Study tip**

Be sure to identify the difference between 'threats', 'causes' and 'effects' of deforestation and make sure that you use each term correctly.

∞ **links**

For further information on Malaysia's rainforests, go to **http://rainforests.mongabay.com/20malaysia.htm**.

For information on deforestation in Brazil, go to **www.mongabay.com/brazil.html**.

# 2.5 Sustainable rainforest management in Malaysia

## National Forest Policy

Widespread logging in Malaysia started after the Second World War due to improvements in technology (e.g. chainsaws, trucks). The government responded by passing the National Forestry Act in 1977. The Act paved the way for sustainable management of Malaysia's rainforests and had the following aims:

- Develop timber processing to increase the profitability of the exported wood and reduce demand for raw timber. The export of low-value raw logs is now banned in most of Malaysia.
- Encourage alternative timber sources (e.g. from rubber trees).
- Increase public awareness of forests.
- Increase research into forestry.
- Involve local communities in forest projects.

One of the main initiatives of the 1977 Act was to introduce a new approach to forest management known as the **Selective Management System** (diagram **A**). This is recognised as one of the most sustainable approaches to tropical forestry management in the world.

> **In this section you will learn**
>
> the concept of sustainable management of tropical rainforests
>
> the range of national and international options for sustainable management.

> **Key term**
>
> **Selective Management System:** a form of sustainable forestry management adopted in Malaysia.

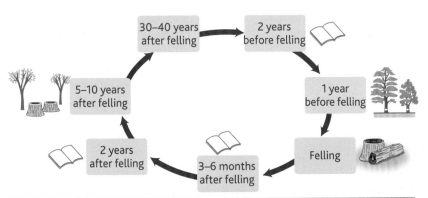

| Stage | Actions |
|---|---|
| 2 years before felling | Pre-felling study to identify what is there. |
| 1 year before felling | Commercially viable trees marked for felling. Arrows painted on trees to indicate direction of felling to avoid damaging other valuable trees. |
| Felling | Felling carried out by licence holders. |
| 3–6 months after felling | Survey to check what has been felled. Prosecution may result from illegal felling. |
| 2 years after felling | Treatment plan drawn up to restore forest. |
| 5–10 years after felling | Remedial and regeneration work carried out by state forestry officials. Replacement trees planted. |
| 30–40 years after felling | Cycle begins again. |

**A** *Malaysia's Selective Management System*

Unfortunately, a lack of trained officials to enforce and monitor the system across the country has led to the continuation of abuses and illegal activities. Remedial measures, such as replanting, have not always been carried out satisfactorily. Deforestation is still taking place in Borneo where land is being converted to oil palm plantations.

## Permanent Forest Estates and National Parks

Land-use surveys carried out in the 1960s and 1970s have enabled the government to identify Permanent Forest Estates. These areas are protected, with no development or conversion of land use allowed. Large areas of forest are, however, used for commercial logging. Some 10 per cent of the forested land (essentially the primary forest areas) has special **conservation** status ensuring the survival of the rainforest habitats and species.

## Forest Stewardship Council

The Forest Stewardship Council (FSC) is an international organisation that promotes sustainable forestry. Products that have been sourced from sustainably managed forests carry the FSC label. The FSC tries to educate manufacturers and consumers about the need to buy wood from sustainable sources. It also aims to reduce demand for rare and valuable tropical hardwoods.

## Developing tourism

In recent years, Malaysia has promoted its forests as destinations for **ecotourism** ('green tourism'). This aims to introduce people to the natural world without causing any environmental damage (photo **B**). The great benefit of ecotourism is that it enables the undisturbed natural environment to create a source of income for local people without it being damaged or destroyed.

> ### Study tip
>
> The key term here is 'sustainable management'. Understand some approaches to sustainable management of rainforests and take time to learn some case study information.

> ### Key terms
>
> **Conservation**: the thoughtful use of resources; managing the landscape in order to protect ecosystems and cultural features.
>
> **Ecotourism**: tourism that focuses on protecting the environment and the local way of life. Also known as green tourism.
>
> **Debt relief**: many poorer countries are in debt, having borrowed money from richer countries to support their economic development. There is strong international pressure for the developed countries to clear these debts – this is debt relief.
>
> **Debt**: money owed to others, to a bank or to a global organisation such as the World Bank.

**B** *Ecotourism in Borneo*

**Features of ecotourism**

- Usually involves small groups.
- Local guides used.
- Buildings use local materials and are environmentally friendly (sustainable water, energy and waste management). Their construction and maintenance provides employment for local people.
- Mostly nature-based experiences (walks, birdwatching).
- Limited transport involved.

# Recent worldwide initiatives

Rainforests are valuable resources particularly for poor countries wishing to expand their economies. Apart from the timber itself, rainforests occupy land that could be used for commercial agriculture such as plantations (e.g. Malaysia) or ranching (e.g. Brazil). Valuable mineral resources such as bauxite, copper or iron may be present in the rock beneath the forests. To expect countries to 'mothball' their rainforests is naïve, however important they are at the global scale.

## Debt relief

One approach is to recognise the international importance of rainforests by giving them a monetary value and paying countries to maintain them. This could take the form of **debt relief**, for example, where countries are relieved of some of their **debt** in return for retaining their rainforests.

**C**   *Scientists in Sierra Leone's Gola Forest*

## Carbon sinks

In 2008 the Gola Forest on Sierra Leone's southern border with Liberia was protected from further deforestation by becoming a National Park (photo **C**). In recognition of the forest's role in reducing global warming by acting as a **carbon sink**, the 75,000 ha park is supported by money from the European Commission, the French government and **non-governmental organisations (NGOs)** such as the Royal Society for the Protection of Birds (RSPB) and Conservation International.

### Activities

1   Study diagram **A**.

a   Make a copy of the diagram showing the Selective Management System.

b   Use the information in the table to add more detail in the boxes in your diagram.

c   Why is the system a good example of sustainable management?

d   Can you suggest any modifications that might make it even more sustainable?

e   Suggest some possible problems that might exist in trying to implement this approach to forest management.

2   Study photo **B**.

a   What is meant by ecotourism?

b   What aspects of ecotourism are evident in the photo?

c   How does ecotourism offer opportunities to protect rainforests from deforestation?

d   Ecotourism has become popular and travel companies are keen to attach this label to many tours. How might this trend cause problems in the future for Borneo's rainforests?

e   Use the internet to look for an example of an ecotourism trip to Borneo. Describe the nature of the trip with the aid of photos. Assess whether it is genuinely ecotourism.

### Key terms

**Carbon sink**: forests are carbon sinks because trees absorb carbon dioxide from the atmosphere. They help to address the problem of global carbon emissions.

**Non-governmental organisation (NGO)**: an organisation that collects money and distributes it to needy causes, e.g. Oxfam, ActionAid and WaterAid.

### links

For further information on the Forest Stewardship Council, visit **www.fsc.org/77.html**.

More details about a National Park, Taman Negara in Malaysia, can be found at **www.geographia.com/malaysia/taman.html**.

Ecotourism information can be found at **www.about-malaysia.com/adventure/eco-tourism.htm**.

# 2.6    What are the opportunities for economic developments in hot deserts?

## In this section you will learn

the economic opportunities of deserts in more economically developed countries (MEDC) and less economically developed countries (LEDC)

the challenges faced by desert communities and the management responses.

**Case study**

## The Thar Desert, Rajasthan, India (an LEDC)

The Thar Desert is one of the major hot deserts of the world. It stretches across north-west India and into Pakistan . The desert covers an area of some 200,000 km², mostly in the Indian state of Rajasthan.

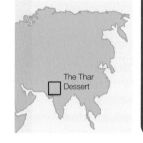

The Thar Dessert

Rainfall in the Thar Desert is low – typically between 120 and 240 mm per year – and summer temperatures in July can reach 53°C. Much of the desert is sandy hills with extensive mobile sand dunes and clumps of thorn forest vegetation, a mixture of small trees, shrubs and grasses (photo **A**). The soils are generally sandy and not very fertile, as there is little organic matter to enrich them. They drain quickly so there is little surface water.

### Economic opportunities in the desert

#### Subsistence farming

Most of the people living in the desert are involved in farming. The climate presents huge challenges, with unreliable rainfall and frequent droughts. The most successful basic farming systems involve keeping a few animals on the grassy areas and cultivating vegetables and fruit trees. Although a good deal of the farming is **subsistence farming**, some crops are sold at local markets.

Over the border in Pakistan's Thar region, the Kohlis tribe are descendants of **hunter-gatherers** who survived in the desert by hunting animals and gathering fruit and natural products such as honey. This type of subsistence farming is the most basic form of farming and is rarely found in the world today.

#### Irrigation and commercial farming

Irrigation in parts of the Thar Desert has revolutionised farming in the area. The main form of irrigation in the desert is the Indira Gandhi (Rajasthan) Canal. The canal was constructed in 1958 and has a total length of 650 km. Two of the main areas to benefit are centred on the cities of Jodhpur and Jaisalmer, where over 3,500 km² of land is under irrigation. **Commercial farming** in the form of crops such as wheat and cotton now flourishes in an area that used to be scrub desert. The canal also provides drinking water to many people in the desert.

#### Mining and industry

The state of Rajasthan is rich in minerals. The desert region has valuable reserves of gypsum (used in making plaster for the construction industry and in making cement), feldspar (used to make ceramics), phospherite (used for making fertiliser) and kaolin (used as a whitener in paper).

There are valuable reserves of stone in the area. At Jaisalmer the Sanu limestone is the main source of limestone for India's steel industry. Limestone is also quarried for making cement. Valuable reserves of the rock marble are quarried near Jodhpur for use in the construction industry. Local hide and wool industries form a ready market for the livestock that are reared in the area.

**links**

Some excellent maps are available at www.mapsofindia.com/geography and www.mapsofindia.com/maps/rajasthan.

**Key terms**

**Subsistence farming**: farming to produce food for the farmer and his/her family only.

**Hunter-gatherers**: people who carry out a basic form of subsistence farming by hunting animals and gathering fruit and nuts.

**Commercial farming**: a type of farming where crops and/or livestock are sold to make a profit.

**Did you know** ??????

The Thar Desert National Park in India is home to the rare great Indian bustard, a large ground-dwelling bird.

**A**  *The natural desert environment near Jaisalmer*

*Tourism*

In the last few years, the Thar Desert, with its beautiful landscapes, has become a popular tourist destination. Desert safaris on camels, based at Jaisalmer, have become particularly popular with foreigners as well as wealthy Indians from elsewhere in the country. Local people benefit by acting as guides or by rearing and looking after camels.

## Future challenges

The Thar Desert faces a number of challenges for the future:

- Population pressure – the Thar Desert is the most densely populated desert in the world, with a population density of 83 people per km², and the population is increasing. This is putting extra pressure on the fragile desert ecosystem and leading to overgrazing and overcultivation.

- Water management – excessive irrigation in some places has led to waterlogging of the ground. Where this has happened, salts poisonous to plants have been deposited on the ground surface. This is called **salinisation** and is a big problem in deserts (diagram **B**). Elsewhere, excessive demand for water has caused an unsustainable fall in water tables.

- Soil erosion –overcultivation and overgrazing have damaged the vegetation in places, leading to soil erosion by wind and rain. Once eroded away, the soil takes thousands of years to re-form.

- Fuel – reserves of firewood, the main source of fuel, are dwindling with the result that people are using manure as fuel rather than using it to improve the quality of the soil.

- Tourism – although tourists bring benefits such as employment and extra incomes, the environment that they have come to enjoy is fragile and will suffer if tourism becomes overdeveloped.

**Salinisation**: the deposition of solid salts on the ground surface following the evaporation of water. Also an increase in the concentration of salts in the soil, reducing fertility.

## Sustainable management

A number of approaches have been adopted to address the challenges of living in the Thar Desert and to provide its people with a sustainable future. In 1977 the government-funded Desert Development Programme was started. Its main aims are to restore the ecological balance of the region by conserving, developing and harnessing land, water, livestock and human resources. In Rajasthan it has been particularly concerned with developing forestry and addressing the issue of sand dune stabilisation.

## Forestry

The most important tree in the Thar Desert is the *Prosopis cineraria*. It is extremely well suited to the hostile conditions of the desert and has multiple uses (photo **C**). Scientists at the Central Arid Zone Research Institute have developed a hardy breed of plum tree called a Ber tree. It produces large fruits and can survive in low rainfall conditions. The fruits can be sold and there is the potential to make a decent profit.

## Stabilising sand dunes

The sand dunes in the Thar Desert are very mobile. In some areas they form a threat to farmland, roads and waterways. Various approaches have been adopted to stabilise the sand dunes, including planting blocks of trees and establishing shelterbelts of trees and fences alongside roads and canals.

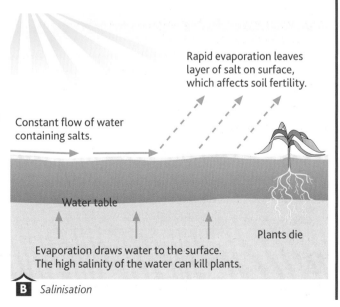

Rapid evaporation leaves layer of salt on surface, which affects soil fertility.

Constant flow of water containing salts.

Water table

Plants die

Evaporation draws water to the surface. The high salinity of the water can kill plants.

**B**  *Salinisation*

## Thar Desert National Park

The Thar Desert National Park has been created to protect some 3,000 km² of this arid land and the endangered and rare wildlife that has adapted to its extreme conditions.

■ A lot of foliage is produced, which can be used to feed animals, especially in the drier winter.

■ The trees can provide good-quality firewood.

■ The wood is strong and can be used as a local building material.

■ Its pods provide animal fodder.

■ Crops can benefit from shade and moist growing conditions if interspersed between the trees.

■ This and other tree species are planted in blocks to help stabilise the sand dunes.

**C**  Sustainable qualities of the Pros-opis cineraria tree

### ⚭ links

Information about the Sonoran Desert can be accessed at **http://alic.arid.arizona.edu/sonoran.**

## Activities

1  Study photo **A**.

a  Describe the environment of the Thar Desert as shown in the photo.

b  What are the challenges of this environment for local people?

c  What is the difference between subsistence farming and commercial farming?

d  Describe the characteristics of these two types of farming in the Thar Desert.

e  Apart from farming, what other economic activities take place in the desert?

2  Study diagram **B**.

a  With the aid of a diagram, describe the process of salinisation.

b  Why does salinisation occur in the Thar Desert?

c  Why is salinisation a problem for the future?

d  Suggest ways of reducing the problem of salinisation.

e  Apart from salinisation, what other challenges face the people of the Thar Desert in the future?

3  Study the information in figure **C**. Apart from stabilising sand dunes, what other benefits do trees provide?

4  a  What are the benefits of sand dune stabilisation?

b  Is sand dune stabilisation a form of sustainable management? Explain your answer.

## The Sonoran Desert, Arizona, USA (an MEDC)

**D**  Location map of the Sonoran Desert

The Sonoran Desert is one of North America's largest and hottest deserts. It is also one of the wettest, with over 300 mm of rain falling in some places. It is located in the south-west of the USA, straddling the lower states of Arizona and California and stretching south into Mexico (map **D**). The Sonoran Desert is stunningly beautiful and is home to a great diversity of flora and fauna including the iconic saguaro cactus (fact file **H**, page 41).

The USA is able to respond somewhat differently to the challenges and opportunities of a desert environment compared with poorer countries such as India (Thar Desert) or the African countries bordering the Sahara Desert. Money enables many of the physical difficulties to be overcome.

The physical extremes of the climate can be overcome to some extent by using air conditioning for vehicles, houses, workplaces and shopping centres. With plentiful supplies of relatively cheap energy, this is perfectly possible in the USA.

Water can be relatively easily piped into the area for irrigating crops, to supply drinking water and for filling swimming pools and watering golf courses.

The clear, clean atmosphere and open spaces form an attraction to short-term holidaymakers and long-term migrants. A recent trend in the Sonoran Desert has been **retirement migration**, where people decide to retire to newly built housing complexes with swimming pools and golf courses.

## Marana: the tale of one town in the Sonoran Desert

Marana is a town of some 35,000 people located a few kilometres north-west of the city of Tucson in Arizona. Over the years it has developed into a thriving business town and leisure resort.

The town began as a mid-19th century ranching and mining community along the Southern Pacific Railroad. In 1920 a new irrigation system enabled it to become an agricultural centre specialising in cotton, a crop that does well in hot conditions provided it is well watered. Families migrated to the town to work in the cotton fields. Agricultural production increased during the 1940s and expanded to include wheat, barley and pecans.

However, since the 1990s farming in the area has declined. In 2005 only six large cotton farms remained. Durum wheat is grown and exported to Italy to make pasta. A heritage park has been opened to celebrate the town's agricultural heritage.

Migration accounts for much of the growth of the town, which is a thriving and wealthy business community.

In 2007 Marana began hosting golf's PGA Matchplay Championship (photo **E**).

> **Key term**
>
> **Retirement migration**: migration to an area for retirement.

## Managing the Sonoran Desert

In 1998 the Sonoran Desert Conservation Plan was initiated in Pima County, the administrative region incorporating Tucson in south-west Arizona. This is a comprehensive plan to 'conserve the county's most valued natural and cultural resources, whilst accommodating the inevitable population growth and economic expansion of the community'.

The plan resulted from concern about threats to wildlife habitats as housing developments expanded into the desert. An endangered species of pygmy owl was considered to be particularly vulnerable.

Among other initiatives, the plan has led to:

- detailed mapping and inventory of the county's natural and cultural heritage
- development of buffer zones around areas of ecological significance
- native plant protection
- hillside development restrictions
- home design recommendations to conserve energy and water.

**E** *Golf course in the desert near Marana*

**Activities**

5  Study photo **E** and the information in the text about Marana.

a  Describe the characteristics and location of Marana using the photo to help you.

b  Draw a timeline to describe the development of Marana since the mid-19th century.

6  What is retirement migration and why is it a significant issue in Arizona?

7  What are the current challenges facing the planners in Marana and how does the Sonoran Desert Conservation Plan address some of these issues?

8  To what extent do you think richer countries such as the USA are better able to exploit the opportunities and address the challenges of desert environments than poorer countries such as India?

# 3 River processes and pressures

## 3.1 How and why do river valleys change downstream?

A river valley is subjected to the three main landscape-shaping processes of **erosion**, **transportation** and **deposition**. It is the extent to which they occur and where they dominate that is critical in shaping the valley.

### Processes of erosion

A river near to its source concentrates on erosion, and especially downward erosion. Photo **A** of Golden Clough, Edale, shows one such river **channel** and its valley. There are four ways in which a river erodes. These are **hydraulic action**, **abrasion**, **attrition** and **solution**.

- Hydraulic action is the sheer force of the water hitting the bed and the banks. This is most effective when the water is moving fast and there is a lot of it.

- Abrasion occurs when the **load** the river is carrying repeatedly hits the river bed and the banks, causing some of the material to break off.

- Attrition is when the stones and boulders carried by the river knock against each other and over time are weakened, causing bits to fall off and reduce in size.

- Solution occurs only when the river flows on certain types of rock, such as chalk and limestone. These are soluble in rainwater and become part of the water as they are dissolved by it.

Rivers tend to erode in one of two directions: downwards or sideways. The terms for these are **vertical** and **lateral erosion**. As a river gets further down its course, vertical erosion becomes less important and lateral erosion takes over.

### Processes of transportation

Having been successful in dislodging parts of the bed and banks, the river then transports the load downstream. There are four methods of transportation. These are **traction**, **saltation**, **suspension** and **solution** (diagram **B**). Photos **C** and **D** show how a river and its valley changes with distance downstream.

<table>
<tr><td>

**In this section you will learn**

the processes of erosion: hydraulic action, abrasion, attrition and solution

the processes of transportation: traction, saltation, suspension and solution

how and why deposition occurs

how and why the long and cross profiles are formed and why cross profiles change downstream.

</td></tr>
</table>

**A** Golden Clough, Edale

**Study tip**

Learn the process terms for erosion and transportation and be certain you use each term in its correct context.

**Traction** is the method used for moving the largest material. This is too heavy to lose contact with the bed, so material such as boulders is rolled along.

**Saltation** moves the small stones and grains of sand by bouncing them along the bed. This lighter load leaves the river bed in a hopping motion.

**Suspension** is a means of carrying very fine material within the water, so that it floats in the river and is moved as it flows.

**Solution** is the dissolved load and occurs only with certain rock types that are soluble in rainwater. This is true of chalk and limestone and the load is not visible.

 **B** *Processes of transportation*

**C** *Grindsbrook Clough, Edale*

**D** *River Noe, Edale*

## Key terms

**Erosion**: the sculpting of a landscape, for example by rivers, involving the removal of material.

**Transportation**: the carrying of sediment downstream from the point where it has been eroded to where it is deposited.

**Deposition**: the dumping (deposition) of sediment that has been transported by a river.

**Channel**: the part of the river valley occupied by the water itself.

**Hydraulic action**: the power of the volume of water moving in the river.

**Abrasion**: happens when larger loads carried by the river hits the bed and banks, causing bits to break off.

**Attrition**: the knocking together of stones and boulders, making them gradually smaller and smoother.

**Solution**: the dissolving of rocks and minerals by rainwater. This is a means of transportation as well as an erosion process.

**Load**: material of any size carried by the river.

**Vertical erosion**: downwards erosion, for example when a river gouges out a deep valley.

**Lateral erosion**: sideways erosion, for example in a river channel at the outside bend of a meander.

**Traction**: the rolling along of the largest rocks and boulders.

**Saltation**: the bouncing movement of small stones and grains of sand along the river bed.

**Suspension**: small material carried (suspended) within the water.

## Deposition

This is where the river dumps or leaves behind material that it has been carrying. It deposits the largest material first as this is the heaviest to carry. The smaller the load, the further it can be transported, so this is deposited much further downstream than the larger load. Thus, large boulders can be seen in photo **A**, whereas they are absent in photo **D**. The river drops some of its load when there is a fall in the speed of the water or the amount of water is less. This often occurs when the gradient changes at the foot of a mountain or when a river enters a lake or the sea.

## Long and changing cross profiles

The **long profile** shows how the river changes in height along its course. Diagram **E** shows a theoretical long profile from the source to the mouth. The steep reduction in height near the source gives way to a more gradual reduction further downstream, giving a typical concave profile. The river has much potential energy near the source due to the steep drop. Later on, this is replaced by energy from a large volume of water. However, such a perfect long profile is rare. This is due to land being uplifted, sea level changing and bands of hard and soft rock crossing the path of the river. As the river flows downstream, its valley changes shape and the **cross profile** from one side of the valley to the other clearly shows this. Generally, the cross profile shows the valley becoming wider and flatter, with lower valley sides. Map extract **F** shows part of the course of the River Noe and some of its tributaries, where photos **A**, **C** and **D** were taken. This shows how the valley cross profile changes with distance from the source.

## ⬭ links

You can find out more facts about large rivers at **http://ga.water.usgs.gov/edu/riversofworld.html** and **http://en.wikipedia.org/wiki/Rivers_of_England**. On this website, it would be worth searching for Durham and then the Tees as preparation for the next section.

### Key terms

**Long profile**: a line representing the course of the river from its source (relatively high up) to its mouth where it ends, usually in a lake or the sea, and the changes in height along its course.

**Cross profile**: a line that represents what it would be like to walk from one side of a valley, across the channel and up the other side.

### Activities

1. Study photo **A**. Draw a labelled sketch to show the characteristics of the channel and valley of Golden Clough. Include comments about the following in your labels:
   - width/depth of channel
   - size of load
   - profile of bed and banks
   - what water flow is like
   - valley sides.

2. For the terms relating to erosion, produce an illustrated dictionary to give your own clear definitions, supported by a simple diagram.

3. For the four types of transportation, produce a diagram to illustrate all four processes. Write a definition of your own next to the relevant part of the diagram.

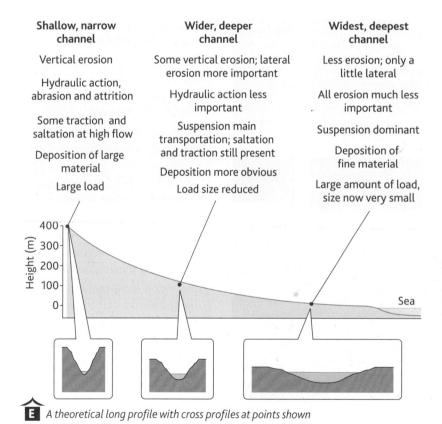

| Shallow, narrow channel | Wider, deeper channel | Widest, deepest channel |
|---|---|---|
| Vertical erosion | Some vertical erosion; lateral erosion more important | Less erosion; only a little lateral |
| Hydraulic action, abrasion and attrition | Hydraulic action less important | All erosion much less important |
| Some traction and saltation at high flow | Suspension main transportation; saltation and traction still present | Suspension dominant |
| Deposition of large material | Deposition more obvious | Deposition of fine material |
| Large load | Load size reduced | Large amount of load, size now very small |

**E** *A theoretical long profile with cross profiles at points shown*

## Activities

**4** Study photos **A**, **C** and **D** and diagram **B**.

a How do you think the load shown in photo **A** will be moved? Explain why.

b What process of transportation is likely to be taking place in photo **C**? Give reasons for choosing the process you have selected.

c Describe how the channel and valleys of the two rivers shown in photos **A** and **C** have changed.

d Describe further changes that have occurred between photos **C** and **D**.

**5** Study diagram **E**.

a Draw a sketch long profile of a river.

b Mark with a dot where you think the photos in **A** and **D** could have been taken.

c Label your long profile to show how and why the size of deposited material changes downstream.

**6** Study map extract **F**.

a Give the approximate height of the rivers where each photo (**A**, **C** and **D**) was taken.

b Describe the gradient of the long profile at each of the locations where photos **A**, **C** and **D** were taken.

c Draw a sketch of the cross profile for each of the three locations.

d Describe how the cross profiles change downstream.

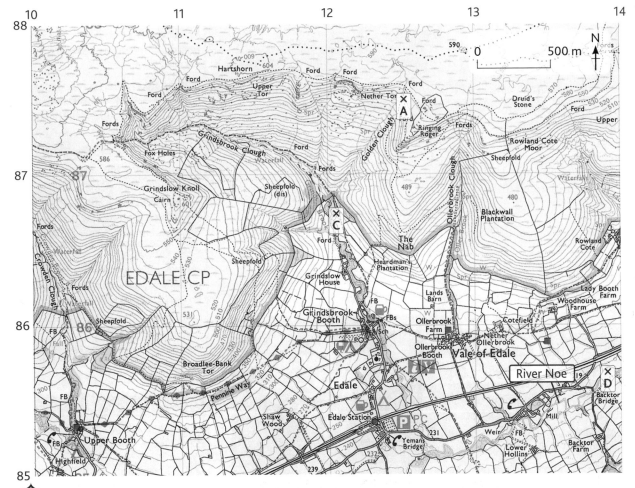

**F** 1:25,000 OS map extract of the River Noe and its tributaries, part of map sheet OL1

# 3.2 What distinctive landforms result from the changing river processes?

Different river processes lead to different landforms. Therefore, in areas near the source where vertical erosion is dominant, **waterfalls** and **gorges** are characteristic features. Further down, where lateral erosion and deposition become more important, **meanders** and **oxbow lakes** develop. Nearer to the mouth, where deposition is the most significant process, **floodplains** and **levees** become a key aspect of the landscape. The River Tees will be used here to illustrate different landforms and how they change downstream.

## ▍Landforms resulting from erosion: waterfalls and gorges

Waterfalls provide some of the most spectacular scenery in mountainous areas. In their wake, they leave gorges as they retreat back up the valley. Diagram **A** shows the sequence of events that occurs in waterfall formation.

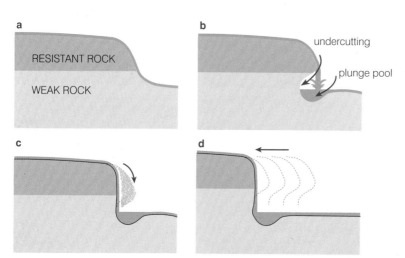

**A** The formation of waterfalls and gorges

One of the best-known waterfalls in the UK is High Force on the River Tees (photo **B** and map extract **C**). This occurs where whinstone, a resistant igneous rock, overlays softer limestone. Water cascading over the waterfall has formed a deep plunge pool.

**B** High Force, River Tees

### In this section you will learn

| |
|---|
| how and why waterfalls and gorges form due to erosion |
| the formation of meanders and oxbow lakes due to erosion and deposition |
| the development of levees and floodplains, due to deposition |
| how the formation of some features is linked to the development of others. |

### Key terms

**Waterfall**: the sudden, and often vertical, drop of a river along its course.

**Gorge**: a narrow, steep-sided valley.

**Meander**: a bend or curve in the river channel.

**Oxbow lake**: a horseshoe or semi-circular area that used to be a meander. Oxbow lakes are cut off from a supply of water and so will eventually become dry.

**Floodplain**: the flat area next to the river channel, especially in the lower part of the course. This is a natural area for water to spill onto when the river reaches the top of its banks.

**Levees**: raised banks along the course of a river in its lower course. They are formed naturally but can be artificially increased in height.

### Did you know ??????

Eas a'Chual Aluinn in Scotland is the UK's highest waterfall at 200 m. This is relatively small when compared with the highest waterfall in the world, Angel Falls in Venezuela at a height of 979 m.

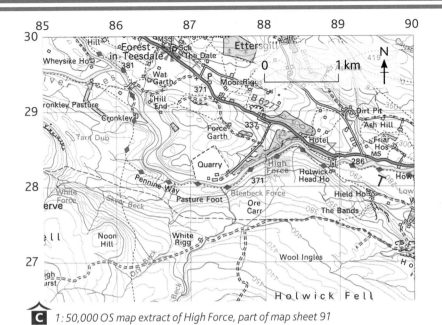

**C** 1 : 50,000 OS map extract of High Force, part of map sheet 91

## Landforms resulting from erosion and deposition: meanders and oxbow lakes

Meanders and oxbow lakes are characteristic landforms in the middle part of the river. The formation of meanders leads eventually to the development of oxbow lakes, as shown in diagram **D**.

Map extract **E** shows meanders further down the course of the Tees. Can you suggest where an oxbow lake may form?

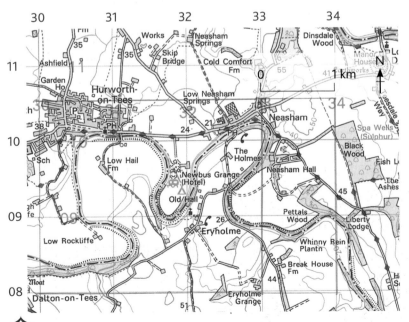

**E** 1 : 50,000 OS map extract of the River Tees south-east of Huxworth-on-Tees, part of map sheet 93

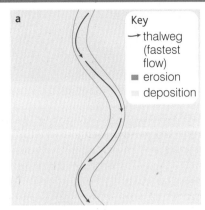

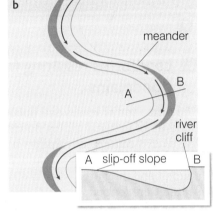

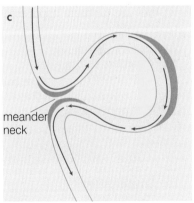

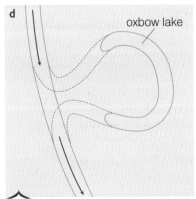

**D** The formation of meanders and oxbow lakes

**F**   *Inside bend of a meander*

## Landforms resulting from deposition: levees and floodplains

The formation of levees and floodplains are linked and involve the build-up of material during flooding. Under normal flow conditions, the river is contained within its banks and so no sediment is available to form levees or the floodplain. However, during periods of high rainfall and discharge when the river has burst its banks, both of these features are formed (photo **G** and diagram **H**). Map extract **J** shows the area much nearer the mouth at Thornaby-on-Tees. Levees are present here; they are far more apparent in photo **I**.

**G**   *Extensive flooding at Barje near Ljubljana, Slovenia*

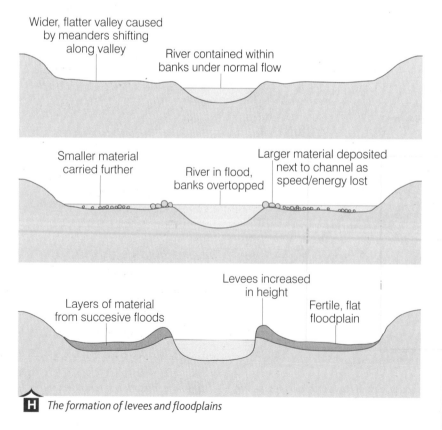

Wider, flatter valley caused by meanders shifting along valley

River contained within banks under normal flow

Smaller material carried further

River in flood, banks overtopped

Larger material deposited next to channel as speed/energy lost

Layers of material from succesive floods

Levees increased in height

Fertile, flat floodplain

**H**   *The formation of levees and floodplains*

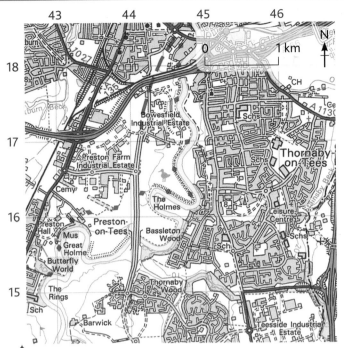

**J** *1 : 50,000 OS map extract of the floodplain of the River Tees, part of map sheet 93*

**I** *Artificial levees on the Mississippi*

**Did you know** ??????

The Mississippi River is 3,800 km long. It carries on average 42,002 tonnes of sediment each day and 130 million tonnes a year.

∞**links**

You can find out more about the River Tees at **http://en.wikipedia. org/wiki/River_Tees**.

## Activities

1  Study diagram **A**. Write a series of bullet points to describe the sequence of waterfall formation.

2  Study photo **B** and map extract **C**.

 a  Research High Force waterfall using the links suggested to obtain specific facts about the waterfall.

 b  Produce an information board about High Force. Include on your board a labelled sketch of the waterfall to describe its features and some facts about it that will interest visiting tourists.

 c  Describe the channel and the valley of the River Tees shown in the map extract.

3  Study diagram **D**.

 a  Draw simplified copies of the diagrams.

 b  Label each diagram to show the stages in the formation of meanders and oxbow lakes.

4  Study map extract **E**.

 a  Draw a sketch map of the meanders showing:

 ■ an inside bend

 ■ an outside bend

 ■ the neck of a meander

 ■ the meander most likely to be cut off first.

 b  Describe additional information about the meanders that is present on the map.

 c  Locate Low Hail Farm (309097) and The Holmes (325098). Describe the location of the two farms and suggest why they are located here.

 d  Comment on the risk of these two farms being flooded by the river.

5  Study photo **F**.

 a  Would you expect the deepest water in the river to be on the left or the right side? Explain your answer.

 b  Do you think this photo was taken at low flow or during a flood? Justify your answer.

 c  What evidence is there that the river rarely flows at a high level?

 d  Draw a sketch of the meander and add labels to identify the main landforms, the different river depths and the line of fastest flow.

6  Study diagram **H** and photo **I**.

 a  Working in pairs, produce a short PowerPoint presentation to include the following:

 ■ the formation of levees

 ■ the formation of floodplains

 ■ the links between the two landforms.

 b  Show your presentation to another group in the class.

# 3.3  How and why does the water in a river fluctuate?

**In this section you will learn**

how the amount of water in a river varies

the reasons for this variation

how hydrographs can be used to show the link between rainfall and discharge.

The **discharge** of a river is the volume of water passing down a river. It can fluctuate a lot in a matter of hours in response to periods of rain. An understanding of the **drainage basin** hydrological cycle (diagram **A**) is useful background in explaining how and why the amount of water in a river is variable.

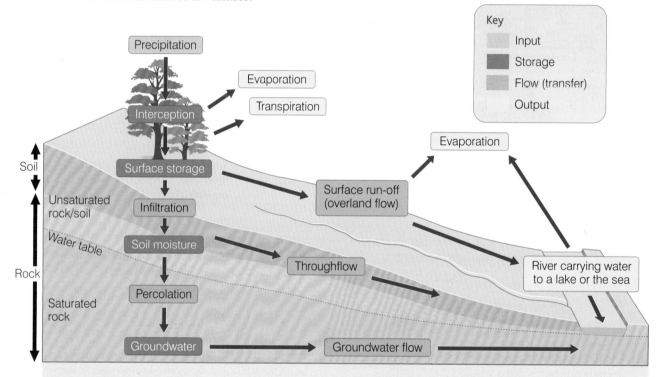

Key
- Input
- Storage
- Flow (transfer)
- Output

**Precipitation**: any source of moisture reaching the ground, e.g. rain, snow, frost

**Interception**: water being prevented from reaching the surface by trees or grass

**Surface storage**: water held on the ground surface, e.g. puddles

**Infiltration**: water sinking into soil/rock from the ground surface

**Soil moisture**: water held in the soil layer

**Percolation**: water seeping deeper below the surface

**Groundwater**: water stored in the rock

**Transpiration**: water lost through pores in vegetation

**Evaporation**: water lost from ground/vegetation surface

**Surface run-off (overland flow)**: water flowing on top of the ground

**Throughflow**: water flowing through the soil layer parallel to the surface

**Groundwater flow**: water flowing through the rock layer parallel to the surface

**Water table**: current upper level of saturated rock/soil where no more water can be absorbed

**A**  Drainage basin hydrological cycle

## The storm hydrograph

The **flood or storm hydrograph** is used to show how a river responds to a period of rainfall (graph **C**). Rivers that respond rapidly to rainfall have a high peak and short lag time and are referred to as **flashy**. A lower peak and long lag time shows a delayed hydrograph. Table **B** gives discharge data for the River Eden in Carlisle between 7 and 9 January 2005.

**Did you know** ??????

Some places in England received over 380 mm of rain in May, June and July 2007 – more than double the average for the three months. This exceeded the previous high of 349 mm established in 1789.

**B** *Discharge of River Eden, Carlisle*

| Date | Time | Discharge in cubic metres per second (cumecs) |
|------|------|-----------------------------------------------|
| 7 January | 0000 hours | 90 |
| | 1200 hours | 130 |
| 8 January | 0000 hours | 820 |
| | 1200 hours | 1,400 |
| | 1500 hours | 1,520 |
| 9 January | 0000 hours | 1,000 |
| | 1200 hours | 430 |

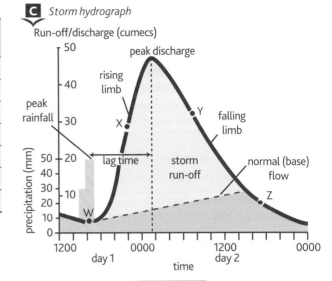

**C** *Storm hydrograph*

## Factors affecting river discharge

River discharge is influenced by a number of factors related to the weather (e.g. rainfall, temperature and previous weather conditions), other physical factors (e.g. **relief** and rock type) and by human land use.

The amount and type of rainfall are important factors. A lot of rain falling causes high river levels, while less rainfall results in lower river levels. The explanation for this lies in the drainage basin hydrological cycle. High amounts of rain saturate the soil and underlying rock. In the case of drizzle, there is time for water to infiltrate the soil and underlying rock, freeing up space for more rain.

Temperature affects the loss of water from the drainage basin and therefore the level of discharge. When temperatures are higher (table **D**), there is greater water loss via evaporation and transpiration, so river levels go down.

**D** *Average rainfall and temperature in Sheffield*

| Month | Rainfall (mm) | Maximum temperature (°C) |
|-------|---------------|--------------------------|
| January | 87 | 6.4 |
| February | 63 | 6.7 |
| March | 68 | 9.3 |
| April | 63 | 11.8 |
| May | 56 | 15.7 |
| June | 68 | 18.3 |
| July | 51 | 20.8 |
| August | 64 | 20.6 |
| September | 64 | 17.3 |
| October | 74 | 13.3 |
| November | 78 | 9.2 |
| December | 92 | 7.2 |

### Key terms

**Discharge:** the volume of water passing a given point in a river at any moment in time.

**Drainage basin:** area from which a river gets its water. The boundary is marked by an imaginary line of highland known as a watershed.

**Flood or storm hydrograph:** a line graph drawn to show the discharge in a river in the aftermath of a period of rain.

**Flashy:** a hydrograph that responds quickly to a period of rain so that it characteristically has a high peak and a short lag time.

**Relief:** height and slope of land.

**Impermeable:** rock that does not allow water to pass through.

**Porous:** rock that allows water to soak into it via spaces between particles.

**Pervious:** rock that allows water to pass through it via vertical joints and horizontal bedding planes.

**Urbanisation:** the increase in the proportion of people living in cities, resulting in their growth.

### Study tip

Ensure that you can explain how physical and human features and processes can affect the amount of water flowing in rivers.

Previous weather conditions also have an impact on river discharge. If it has been dry, it will take longer for the water to reach the river and the amount will be less than if there had been a number of wet days. Table **E** gives the daily totals for Sheffield from 13 June until flooding occurred on the 26th. Notice that a lot of rain had fallen in mid-June before the heavy rain that fell on 24 and 25 June. Steep slopes encourage fast run-off as the water spills rapidly downwards due to gravity.

Rock type is important in determining how much water infiltrates and how much stays on the surface. Areas with **impermeable** rocks, such as granite, have more surface rivers than permeable rocks such as chalk, which is **porous**, and limestone, which is **pervious**.

Land use relates to the function of an area. It is a factor that considers the effect of people. Deforestation (photo **F**) and **urbanisation** (photo **G**) are the two most important land-use changes with regard to influencing river discharge. If trees are removed, water reaches the surface faster and the trees do not extract water from the ground. Expanding towns create an impermeable surface. This is made even worse by building drains to take the water away from buildings quickly – and equally quickly into rivers!

**E**  *Rainfall in Sheffield, June 2007*

| Date | Rainfall (mm) |
| --- | --- |
| 13 June | 30.3 |
| 14 June | 88.2 |
| 15 June | 16.9 |
| 16 June | 0.0 |
| 17 June | 2.2 |
| 18 June | 0.0 |
| 19 June | 14.7 |
| 20 June | 0.1 |
| 21 June | 9.6 |
| 22 June | 6.9 |
| 23 June | 0.2 |
| 24 June | 36.0 |
| 25 June | 51.1 |
| 26 June | 0.0 |

**F**  *Deforestation in Tayside, Scotland*

**G**  *Urbanisation in Leicester*

## Activities

1  Study diagram **A**. Work in pairs to produce a series of flash cards for the key terms in the diagram.

a  On one side of the card, there should be a definition and an illustration of the term.

b  On the other side, the term should be given so that the answer is clear.

c  Use your completed flash cards with another pair to see how many terms you got correct and how good your definitions are.

2  Study table **B** and graph **C**.

a  Draw a line graph to show the discharge of the River Eden in Carlisle.

b  Using the terms in graph **C**, describe the flood hydrograph you have drawn.

c  Can the hydrograph be described as 'flashy'?

3  Study tables **D** and **E** and photos **F** and **G**. Work in pairs to produce an A3 spread around the title (placed in the centre) 'The reasons river discharge fluctuates'.

a  In the central box, give the meaning of 'river discharge'.

b  Place the factors around the central title.

c  Add information to each factor to explain how discharge is affected. Each factor should be illustrated by diagrams, sketches or photos. You should use data provided in the tables.

d  Your finished spread should be informative and accurate but interesting, colourful and original.

# 3.4   Why do rivers flood?

**Floods** occur when a river bursts its banks. Flooding is not a normal condition for the river, but is an extreme situation due to high levels of flow. The extent to which the river exceeds the flow that can be contained in its banks determines the severity of the flood. Floods are common events. Problems and issues arise when people are affected. Building on floodplains results in property being damaged and lives being lost in what becomes a **hazard**.

## ▪ Causes of floods

Rivers flood due to a number of physical causes such as prolonged rainfall, heavy rain, snowmelt and steep relief (table **A** and extracts **B** and **C**). People often unintentionally increase the likelihood and severity of flooding. This is mainly the result of deforestation and construction work.

**A**   *Rainfall in Sheffield*

| Month | Average rainfall (mm), 1971–2000 | Actual rainfall (mm), 2007 |
|---|---|---|
| March | 67.9 | 44.5 |
| April | 62.5 | 5.8 |
| May | 55.5 | 83.8 |
| June | 66.7 | 285.6 |

**In this section you will learn**

why flooding occurs – the natural causes and the ways in which people make it worse

where floods have occurred in the UK

how the frequency of flooding seems to be increasing.

**Key terms**

**Floods**: these occur when a river carries so much water that it cannot be contained by its banks and so it overflows on to surrounding land – its floodplain.

**Hazard**: an event where people's lives and property are threatened and deaths and/or damage result.

**Soil erosion**: the removal of the layer of soil above the rock where plants grow.

---

**News**    Home | Contact | Sitemap | News | Forum | Shop

Home
Contact
Sitemap
News
Forum
Shop

Snowmelt is a main contributing factor to flooding. It was partly responsible for floods in Malton (N. Yorkshire) in November 2000 and is frequently important in Bangladesh. An extract from 'Why Bangladesh floods are so bad' (BBC News website news.bbc.co.uk) on the 2004 floods states 'Bangladesh receives enormous amounts of water from four major rivers. All are filled up from melting snow in the Himalayas.' The Himalayas provide some of the steepest relief worldwide and this is also important in ensuring that water reaches the rivers quickly, increasing the flood risk.

Deforestation has an impact on the water cycle in a drainage basin. It is frequently seen as one reason for the increasing severity of flooding in Bangladesh. Chopping trees down in higher-lying areas, including neighbouring countries, such as Nepal, can have unintentional effects elsewhere. As well as increasing rates of surface runoff, **soil erosion** is a consequence. Much of this is washed into rivers, where, when deposited, the amount of water the channel can hold is reduced. Thus, the flood risk is increased.

Building construction can increase flooding. New houses built next to the river in Malton are clearly susceptible to flooding. Building on floodplains is an issue, as outlined in Lincoln.

**C**   *Causes of flooding*

The city experienced the most rain in a single month since records began 125 years ago. A resident said, 'The rain had been almost constant for a week. On the morning the flooding started, the rivers were almost visibly rising. By lunchtime, the city was at a standstill as bridges became impassable and underpasses flooded. We were stranded in the north of the city – it took six hours to travel four miles. By the evening much of the city centre was under water, many roads had collapsed and were impassable and electricity supplies were limited for almost a week. It took well over a year for things to get back to normal.'

*Adapted from the* Sheffield Star

 *Prolonged rain in Sheffield, 2007*

## Frequency and location of flood events

Flooding appears to be becoming an increasingly frequent event. In 1607 a great flood affected Devon, Somerset and South Wales. Major floods, however, were infrequent in the UK. In March 1947 major floods did occur, affecting many areas of southern, central and north-eastern England, including York, Tewkesbury, Shrewsbury, Sheffield, Nottingham and London, following the rapid melting of snow. The combined effects of storm surge and high tides contributed to the floods of January 1953 that hit the east coast, including Suffolk, Essex and Kent, when huge waves washed away sea defences and 307 people died. In 1968 another Great Flood affected counties in south-east England.

After this, there is little reference to major floods until relatively recently, when they have regularly made the headlines. Since 1998 headlines about floods have been an almost annual occurrence. Table **D** summarises some of the most serious floods since 1998.

**Did you know** ? ? ? ? ?

Brize Norton in Oxfordshire recorded 115 mm of rainfall in 12 hours on 20 July 2007; some places received between 30 and 40 mm in one hour. One unofficial rain gauge in Hull recorded 94 mm.

**links**

You can find out more about causes of flooding at **www.bbc.co.uk**.

**D**  *Major flood events, 1998–2009*

| Date | Location | Rivers |
|------|----------|--------|
| April 1998 | Warwickshire, Gloucestershire, Herefordshire, Worcestershire, Leicestershire, south-east Wales | Avon, Severn, Wye |
| | Northamptonshire, Bedfordshire and Cambridgeshire | Nene, Great Ouse |
| March 1999 | Malton and Norton flooded by Derwent and its tributaries | Derwent |
| May 2000 | Uckfield, Petworth, Robertsbridge, Horsham (Sussex) | Uck, Rother |
| June 2000 | Calder Valley, Yorkshire and York | Calder, Ouse |
| August 2004 | Boscastle (Cornwall) | Valency |
| January 2005 | Carlisle | Eden |
| June 2007 | Large areas of south and east Yorkshire including Doncaster, Sheffield and Hull | Don, Hull, Witham, |
| | Parts of Lincolnshire including Lincoln and Louth | Witham, Ludd |
| July 2007 | Large areas of Gloucestershire including Upton-upon-Severn, Tewkesbury, Gloucester | Avon, Severn |
| | Oxfordshire including Oxford, Banbury and Witney | Thames, Windrush, Cherwell |
| November 2009 | Cockermouth and the Derwent valley to Workington | Derwent, Cocker |

## Activities

1 Study table **A** and extract **B**.

a Draw a comparative bar graph to show the average rainfall between March and June and rainfall that fell in those months in 2007.

b Explain how the figures help to explain the flooding that occurred.

c Consider the comments made by the Sheffield resident in the extract. How do these explain the flooding at the end of June 2007?

2 Study extract **C**.

a Using the example of Bangladesh, explain how snowmelt, relief and deforestation are responsible for flooding. You should include one diagram for each reason.

b Explain why building on floodplains might increase the likelihood of flooding.

c To what extent do you agree with the following statement: 'There is not usually one cause of flooding but a combination of reasons.' Support your answer with evidence.

3 Study table **D**.

a On an outline map of the UK, mark and label the places that flooded.

b Add the major rivers to your map.

c Use internet research to help you add brief facts and figures about each flood as annotations on your map.

# 3.5 How and why do the effects of flooding and the responses to it vary?

The effects of flooding vary according to their size and location. The impact tends to be more severe in LEDCs. Responses are generally more immediate in countries at further stages of development and the attempts made to reduce the effects come from within the affected area or country. In LEDCs attempts made to reduce the effects may be delayed and require international effort. Long-term responses are likely to show similar differences as a result of variations in wealth and the ability to afford flood protection measures.

> **In this section you will learn**
>
> the effects of flooding and the responses to it in both richer and poorer countries
>
> how and why the effects and responses vary.

## Flooding in England (2007)

The flooding of many parts of England in June and July 2007 was the most extensive ever experienced. The depth may not have reached the record levels of 1947, but the scale of the areas affected reached a new high. Diagram **A** shows the main areas affected and the damage done when much of central and southern England suffered the effects of record levels of rainfall.

> **Did you know** ??????
>
> The sale of women's raincoats at John Lewis in July 2007 was over 11 times higher than in July 2006, while the sale of umbrellas was 184 per cent higher.

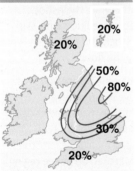

**RISK OF DISRUPTION**
21:00 Sun 24 Jun to
23:00 Mon 25 Jun

20%
20%
20%
50%
80%
30%
20%

- Surface water flooding in Hull.
- Widespread disruption and damage to more than 7,000 houses and 1,300 businesses in Hull.
- River Don burst its banks, flooding Sheffield and Doncaster.
- Flooding in Derbyshire, Lincolnshire and Worcestershire.
- Highest official rainfall total was 111 mm at Fylingdales (North Yorkshire). Amateur networks recorded similar totals in the Hull area.
- There were fears that the dam wall at the Ulley Reservoir near Rotherham would burst.

**B** Flooding in Hull

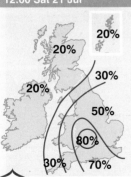

**RISK OF DISRUPTION**
00:00 Fri 20 Jul to
12:00 Sat 21 Jul

20%
20%
30%
20%
50%
80%
30%
70%

- Widespread disruption to the motorway and rail networks.
- In the following days the River Severn and tributaries in Gloucestershire, Worcestershire, Herefordshire and Shropshire broke banks and flooded surrounding areas.
- River Thames and its tributaries in Wiltshire, Oxfordshire, Berkshire and Surrey flooded.
- Flooding in Telford and Wrekin, Staffordshire, Warwickshire and Birmingham.
- The highest recorded rainfall was 157.4 mm in 48 hours at Pershore College (Worcestershire).

**A** Flooding in England, 2007

**C** Victims of the Tewkesbury floods

**Case study**

# Cockermouth, 2009

## Causes of the flood

On 19 November 2009 the highest ever recorded amount of rain in a 24-hour period fell on the hills of the Lake District in NW England. Following a wet autumn, this torrential rain fell on already saturated ground and much of it swept rapidly down the steep hillsides swelling the streams and rivers.

Cockermouth is a small town on the western side of the Lake District 13 km inland from the coast (map **D**). It lies at the confluence of two rivers, the Derwent and the Cocker. The sheer volume of water sweeping down these two rivers overwhelmed the channels and the rivers burst their banks causing widespread and devastating flooding.

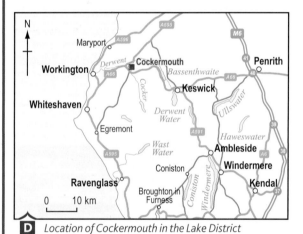

**D**   *Location of Cockermouth in the Lake District*

## Effects of the flood

In Cockermouth itself a torrent of water some 2.5 m high cascaded down the main road flooding and severely damaging shops, offices and homes (photo **E**). In Workington on the coast, a policeman died when the bridge that he had been warning people not to cross collapsed beneath him.

The flood had a big impact on the local community:

- Over 1,300 people were directly affected by the floods.
- Many homes were without power or water and local schools were forced to close.
- All of Cumbria's 1,800 bridges had to be inspected to check that they were safe to use. Several bridges were completely destroyed by the flood.
- In Workington, the northern part of the town became cut off resulting in a lengthy round trip to reach schools, doctors' surgeries and banks. The dislocation of road communications in rural Cumbria caused immense hardship to local people.

- Insurance companies expected the cost of the damage to be in excess of £100m.
- Farmers suffered greatly from the inundation of their land. Livestock were killed and important pastures became useless, either because they were saturated or because they had become coated with rocks carried down by the swollen rivers. Fences and walls were destroyed, farm buildings damaged and machinery ruined.
- Local businesses, such as shops, pubs and hotels faced a dismal winter with little prospect of making any money. Shop stock had been destroyed and many businesses suffered directly from flood damage with up to 1m of mud and silt needing to be cleared from downstairs rooms.

## Responses

Immediately after the flood the main focus was on search and rescue. Some 200 people had to be airlifted from the roofs of their houses by RAF helicopters. People who had been evacuated from their homes were taken to temporary shelters to be looked after. Several people were rescued from their homes by boat.

In the days that followed, flooded buildings were assessed for their safety before residents and shopkeepers were allowed back to assess the damage for themselves. Some small businesses were provided with temporary trading accommodation in the town centre while their shops were refitted.  For days, the streets were lined with skips overflowing with damaged shop stock and people's personal belongings, all ruined by the flood.

Most of the residents, whose homes were flooded by up to 6ft of muddy water, had to live in alternative accommodation for several months. The main road bridge over the River Derwent was closed to vehicles for two months to allow for safety checks and repairs. Several roads in the town centre were so severely broken up by the raging floodwaters that they had to be completely re-surfaced.

**E**   *Flooding in Cockermouth 2009*

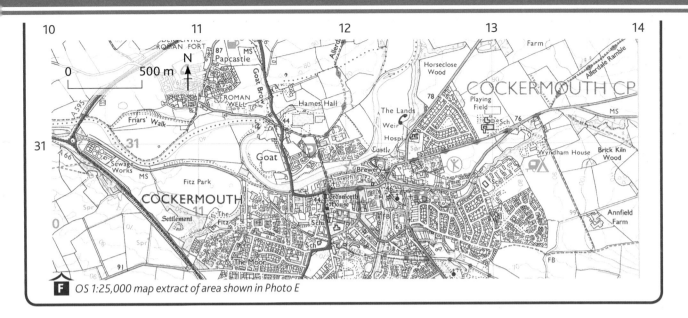

**F** OS 1:25,000 map extract of area shown in Photo E

## Bangladesh, 2004

Bangladesh has experienced particularly severe flooding on a regular basis. Annual flooding is expected in this low-lying country that is located largely on the delta of the Ganges.

The 2004 floods occurred from July to September, inundating over half of the country at their peak. Map **G** shows the areas of Bangladesh under water and the depth of the water.

By July, 40 per cent of Dhaka was under water, 60 per cent of the country was submerged, 600 deaths were reported and 30 million left homeless out of a population of 140 million. Some 100,000 people in Dhaka alone were suffering from diarrhoea as the floodwaters left mud and raw sewage in their wake. As the year progressed, things got even worse as Bangladesh experienced its heaviest rain in 50 years with 35 cm falling in one day on 13 September. The death toll rose to over 750, the airport at Dhaka was flooded as were many roads and railways. Bridges were also destroyed – all of which hampered the relief effort that followed. This damage and that to schools and hospitals was put at $7 billion. In many badly hit rural areas, rice – the main food crop – was washed away and other important food supplies such as vegetables were lost along with cash crops such as jute and sugar.

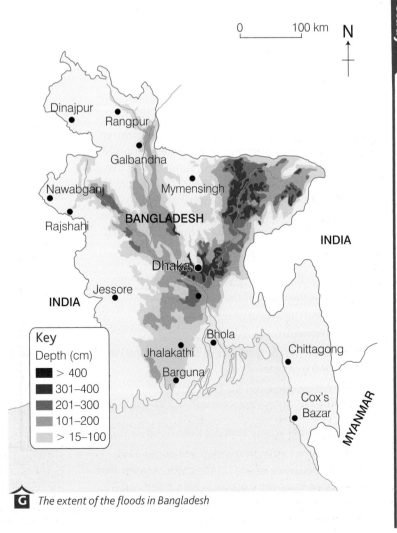

Key
Depth (cm)
■ > 400
■ 301–400
■ 201–300
■ 101–200
■ > 15–100

**G** The extent of the floods in Bangladesh

Within Bangladesh, food supplies, medicines, clothing and blankets were distributed. The effects of the floods on the transport system made this difficult (photo **H**). Local communities began to rebuild their homes. Disease from contaminated water remained a major threat. The United Nations launched an appeal for $74 million, but had received only 20 per cent of this by September. An appeal by WaterAid sought to supply water purification tablets and posters highlighting the hygiene risks in flood water. All of these represent immediate responses to the floods.

Longer-term approaches included embankments built along the river, which have not really achieved their goal, whereas flood warning and the provision of flood shelters have been more successful. These are areas of raised land where people can move to temporarily with their cattle and have access to items such as dried food and obtain water before supplies are contaminated.

**H** *The effects of flooding in rural areas of Bangladesh*

## ⬭⬭ links

You can find out more about flooding in the UK and Bangladesh at **www.bbc.co.uk**.

The website **www.bangladesh.gov.bd** has information about Bangladesh.

## Activities

**1** Study diagram **A**. Imagine you are a newscaster presenting a special report on 'The 2007 floods in England'. Write your script, summarising the location of areas hit and the effects. Make sure that you stress the scale of the flooding and refer to evidence to back up what you are saying. You should have a map to refer to in your broadcast.

**2** Study photo **E** and map **F**.

a  In which direction is the photograph looking?

b  What is the name of the area of land at X?

c  Describe the extent of the flooding in Cockermouth.

d  Describe the likely effects of the flooding on the following people:

■ A shopkeeper on the High Street

■ A homeowner at Y

■ A farmer whose land has been flooded at Z.

e  Locate the castle on the map and photo. How does the choice of site for the castle suggest that flooding has been a problem here in the past?

f  Despite the flooding of 2009 new houses are being constructed on the floodplain at Y. Do you think this is a wise decision?

**3** Study map **G** and photo **H**. Imagine you are working as a volunteer for an aid agency in Bangladesh during the floods of 2004. Write a letter home, describing the situation in the floods in Bangladesh, its effects and responses to it. Try to make your account as real as possible so that the reader can imagine the experience.

**4** Consider the Cockermouth flood (2009) and the Bangladesh flood (2004).

a  Compare the effects and responses to the floods.

b  To what extent do you think the effects and responses reflect the different levels of development of the two countries?

|  | Cockermouth | Bangladesh |
|---|---|---|
| Location |  |  |
| Date |  |  |
| Causes |  |  |
| Effects |  |  |
| Responses |  |  |

# 3.6  Hard and soft engineering: which is the better option?

**Hard engineering** strategies involve the use of technology in order to control rivers. **Soft engineering** adopts a less intrusive form of management, seeking to work alongside natural processes. Hard engineering approaches tend to give immediate results but are expensive. Soft engineering is much cheaper and offers a more sustainable option as it does not interfere directly with the river's flow.

## ▦ Hard engineering

**Dams** and **reservoirs** exert a huge degree of control over a river. Water can be held up behind a dam following heavy rainfall and then released slowly to prevent flooding. Dams and reservoirs are normally constructed as part of a multi-purpose project rather than with just a single aim in mind.

**Straightening meanders** represents a smaller-scale approach to managing rivers. Meanders are circuitous courses. Like following a route in a car, a semicircular way is longer and slower than a straight one. Therefore, water in a meander takes longer to clear an area than water in a straight section of a river. A possible solution to flooding in areas where there are many meanders on a river's course is to straighten them artificially. In this way, the river is made to follow a new shorter, straight section and abandon its natural meandering course (diagram **A**).

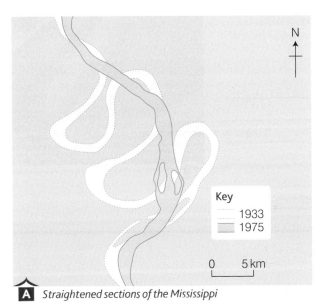

Key
- ┈┈ 1933
- ▬ 1975

0    5 km

**A** *Straightened sections of the Mississippi*

### In this section you will learn

- how hard and soft engineering are used to try to manage rivers and flooding
- why there is debate about the two options
- how to evaluate the two strategies and come to a supported view about them.

### Key terms

**Hard engineering**: building artificial structures aimed at controlling natural processes.

**Soft engineering**: this option tries to work with the natural river system and involves avoiding building on areas especially likely to flood, warning people of a possible flood and planting trees to increase lag time.

**Dam**: an artificial structure designed to hold back water to create a reservoir.

**Reservoir**: commonly an artificial lake formed behind a dam and used for water supply.

**Straightening meanders**: making the river follow a more direct, rather than its natural course, so that it leaves an area more quickly.

### Did you know ??????

The Hoover Dam over the Colorado river near Las Vegas was the biggest concrete structure when it was completed in 1935. There would have been enough concrete to make a two-lane road between San Francisco and New York.

## The Three Gorges Dam, China

The Three Gorges Dam was constructed at Yichang on the River Yangtse (map **B**, photo **C** and table **D**). The capacity of the reservoir should reduce the risk of flooding downstream from a 1-in-10-year event to a 1-in-100-year event. Not only will this benefit over 15 million people living in high-risk flood areas, it will also protect over 25,000 ha of farmland.

The dam is already having a positive impact on flood control, navigation and power generation, but it has caused problems. The Yangtse used to carry over 500 million tonnes of silt every year. Up to 50 per cent of this is now deposited behind the dam, which could quickly reduce the storage capacity of the reservoir.

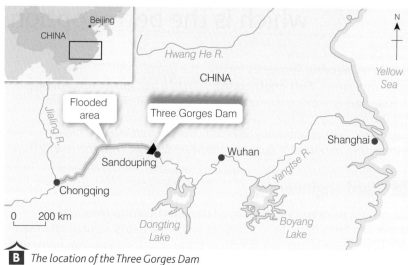

**B**   *The location of the Three Gorges Dam*

The water in the reservoir is becoming heavily polluted from shipping and waste discharged from cities. For example, Chongqing pumps in over 1 billion tonnes of untreated waste per year. Toxic substances from factories, mines and waste tips submerged by the reservoir are also being released into the reservoir.

Most controversially, at least 1.4 million people were forcibly moved from their homes to accommodate the dam, reservoir and power stations. These displaced people were promised compensation for their losses, plus new homes and jobs. Many have not yet received this, and newspaper articles in China have admitted that so far over $30 million of the funds set aside for this has been taken by corrupt local officials.

**D**   *The Three Gorges Dam Project fact file*

| Dimensions | 181 m high and 2.3 km wide |
|---|---|
| Area flooded | 632 km$^2$ |
| Cost | $25.5 billion |
| Built | Started 1994; finished 2009 |
| Increased depth | 110 m (reduced to 80 m when flood risk downstream) |

**C**   *The Three Gorges Dam*

## ■ Soft engineering

Soft engineering seeks to work with the natural processes of a river to reduce the risk of flooding. One approach, that may be appropriate in areas where small floods are frequent, is to do nothing! However, there are many other approaches that can reduce the risk of flooding.

Flood warnings and preparation are complementary approaches. Telling people in advance of a flood gives them valuable time to prepare for it. The Environment Agency identifies areas at risk of flooding and issues warnings. For example, information was given on local radio, television and the internet during the Tewkesbury flooding of 2007. Floodline Warnings Direct sends messages to registered users. A flood watch was issued on 20 July at 18.32, followed by a flood warning approximately 20 minutes later. This gave people valuable time to move possessions, turn off services and take precautions. The Environment Agency's website contains general information on how to prepare for a flood and what to do during and afterwards.

**Floodplain zoning** occurs where the flood risk across different parts of the floodplain is assessed and resulting land use takes this into account. Notice in diagram **E** that the most expensive land uses are located on higher land away from the river.

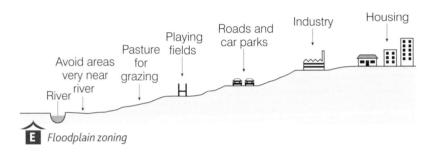

**E** *Floodplain zoning*

In some river basins, deliberate flooding is encouraged on low value land (e.g. marshes or grassland). This prevents flooding of more valuable land downstream.

## ∞ links

You can find out more about the Three Gorges Dam at www.ctg.com.cn

The Environment Agency gives extensive advice about flooding, which you can investigate at www.environment-agency.gov.uk.

### Key terms

**Floodplain zoning**: controlling what is built on the floodplain so that areas that are at risk of flooding have low-value land uses.

**Economic**: relates to costs and finances at a variety of scales, from individuals up to government.

**Social**: refers to people's health, their lifestyle, community, etc.

**Environmental**: the impact on our surroundings, including the land, water and air as well as features of the built-up areas.

## Activities

**1** Study diagram **A**.

a Use the scale to estimate the reduction in the course of the river shown in the diagram after it had been straightened.

b Describe in your own words why river straightening reduces flood risk.

c Suggest some environmental effects of river straightening.

d How might river straightening actually increase the flood risk downstream?

**2** Study map **B**, photo **C** and table **D**.

a Produce a fact file giving six key items of background information on the Three Gorges Dam. Include such things as its location, when it was completed, etc.

b Create a table to give one **economic**, one **social** and one **environmental** cost and benefit of the project.

**3** Access the Environment Agency website from the link above. Working in pairs, produce a leaflet or poster on 'What to do in the event of a flood'. Include information on:

■ warnings available

■ how people can get general and specific information

■ what people should do before, during and after a flood in their home.

Make your leaflet or poster clear, eye-catching, informative and colourful.

**4** Study diagram **E**. Vista Homes has submitted a planning application to build on the area labelled 'Playing fields'. Explain why, as the planner considering the application, you have rejected the scheme.

## 3.7 How are river systems managed to ensure a sustainable water supply?

People in the UK use between 124 and 177 litres of water per day. This is an average of 151 litres per person per day. Table **A** shows actual water use, while the maps in **B** show the current and projected demands in England and Wales. The amount of rainfall over England and Wales varies, as does **water stress** (map **C**). Areas with high water stress can be seen as **areas of water deficit** while those with low stress are often **areas of water surplus**.

**A** *Water use for a selection of activities*

| Activity | Average weekly use | Litres used per activity | Total number of litres |
|---|---|---|---|
| Bath | 2 | 80 | 160 |
| Flushing the toilet | 35 | 8 | 280 |
| Power shower | 7 | 80 | 560 |
| Washing machine | 3 | 65 | 195 |
| Dishwasher | 4 | 25 | 100 |
| Watering the garden | 1 | 540 | 540 |
| Washing car with bucket | 1 | 32 | 32 |
| Washing car with hose pipe | 1 | 450 | 450 |

### *In this section you will learn*

why there is an increasing demand for water in the UK

how there are areas of deficit and areas of surplus and that transfer occurs between them

the economic, social and environmental issues resulting from this transfer.

∞ links

You can find out more about water management at **www.environment-agency.gov.uk**. Click on Business and industry, then Water.

**a** Actual household use, 2005–06

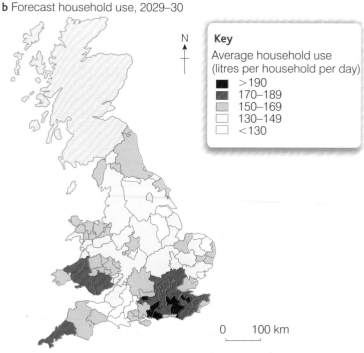

**b** Forecast household use, 2029–30

N

**Key**
Average household use
(litres per household per day)
■ >190
■ 170–189
▨ 150–169
□ 130–149
□ <130

0    100 km

**B** *Household water use in England and Wales*

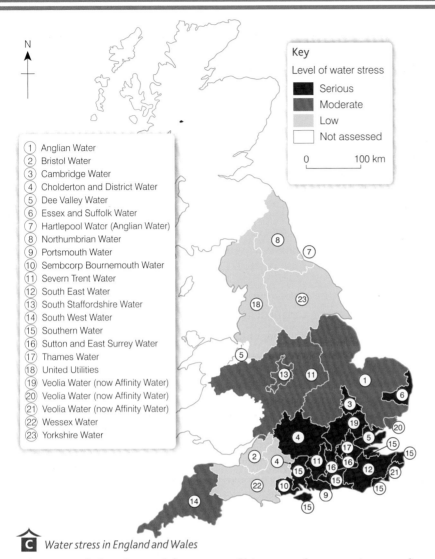

N

Key

Level of water stress

■ Serious
■ Moderate
■ Low
□ Not assessed

0 _____ 100 km

1. Anglian Water
2. Bristol Water
3. Cambridge Water
4. Cholderton and District Water
5. Dee Valley Water
6. Essex and Suffolk Water
7. Hartlepool Water (Anglian Water)
8. Northumbrian Water
9. Portsmouth Water
10. Sembcorp Bournemouth Water
11. Severn Trent Water
12. South East Water
13. South Staffordshire Water
14. South West Water
15. Southern Water
16. Sutton and East Surrey Water
17. Thames Water
18. United Utilities
19. Veolia Water (now Affinity Water)
20. Veolia Water (now Affinity Water)
21. Veolia Water (now Affinity Water)
22. Wessex Water
23. Yorkshire Water

**C** *Water stress in England and Wales*

**Key terms**

**Water stress**: this happens when there is not enough water available. This may be because of an inadequate supply at a particular time or it may relate to water quality.

**Areas of water deficit**: areas where the rain that falls does not provide enough water and there may be shortages.

**Areas of water surplus**: areas that have more water than is needed – often such areas receive a high rainfall total, but have a relatively small population.

**Sustainable**: making sure there is enough water in the long term without harming the environment.

It is expected that demand for water will increase due to an increased number of households and population in certain areas. In 2000, the UK's population was 59.4 million. By 2009 this had risen to 61.8 million. Estimates predict a level of about 66 million by 2031. A more affluent lifestyle increases the demand for water as we buy more time-saving goods. We also demand foodstuffs out of season, which contributes to an increase in overall use of water.

There is a need to ensure that demand can be met in a **sustainable** way. A focus on local schemes is one way of ensuring this, rather than large-scale transfers. Generally, households that have a water meter use less water than those that do not – on average 19 litres per person per day less. When supplies are limited during drought, people use less water. Encouraging conservation is another strategy:

- Houses are being designed with better water efficiency.
- Devices are fitted to toilet cisterns to reduce water use.
- Rainwater can be collected.
- Bath water can be recycled to flush toilets.
- More people are taking showers than baths.

A significant amount of water is also lost in leakage.

**Did you know** ?????

Around 3 per cent of water in the UK is used for drinking, although all water is of drinking quality. Many people in the world survive on less than 10 litres of water a day – a single flush of a toilet uses four-fifths of this.

**Case study**

# Managing River Systems: Kielder Water

Kielder Water in Northumberland was completed in 1982 and is the largest man-made reservoir in northern Europe. It is located about 40 km north west of Newcastle (See Map D). It was built to meet the expected increase in demand for water resulting from a rising population and existence of heavy industry in north-east England. In fact, many of the industries had closed by the time the scheme was built. However, in recent years Kielder Water has come into its own, with underground springs ensuring that the water level remains high, regardless of the weather conditions. This means that while the south of England is often forced to implement drought strategies and hosepipe bans, north-east England enjoys plentiful water supplies. Towns and cities that benefit from the scheme include Newcastle, Sunderland, Middlesbrough, Darlington and Durham. Kielder Water is a multi-purpose reservoir, which means that it not only helps to provide a reliable water supply, but is also used for other purposes such as generating hydroelectric power, flood control, conservation and recreation.

## Why was this site chosen for a dam and reservoir?

- The limited agricultural value of the area, and the small numbers of people living there.
- The natural narrowing of the valley, with steep sides, suitable to build the dam.
- A relatively flat upstream gradient, allowing a maximum volume of water to collect there.
- A band of impermeable volcanic rock (Whinsill) providing a good foundation.
- The large water catchment area.
- Extensive local deposits of boulder clay, sand and gravel needed in the dam construction.
- Heavy and reliable rainfall of 1,400 mm a year.

The scheme involved two stages:

- Building a dam and creating Kielder Water reservoir.
- Pumping water through huge pipes 8 km uphill from the River Tyne to a small holding reservoir. From here the water is transferred 40 km to the River Wear and River Tees.

## What are the advantages and disadvantages of the scheme?

- The dam regulates the flow of the River Tyne, reducing the flow downstream and releasing water during periods of low discharge.
- A large area of land was flooded to create the reservoir, so some farmers lost their livelihoods.

| Factfile: | |
| --- | --- |
| Construction period | 7 years |
| Height of dam | 50 m |
| Length of dam | 1.2 km |
| Materials used | 5 million cubic metres of clay |
| Surface area of Kielder Water | 1084 hectares |
| Maximum supply capacity | 909 million litres per day |
| Volume of water in reservoir | 190 billion litres |
| Length of reservoir | 11 km |
| Maximum depth of water | 52 m |
| Length of lake shoreline | 43 km |
| Number of trees felled | 1.5 million |
| Families rehoused | 7 |
| Cost of scheme | £185m |
| Value of tourism | £6m per year |

- 58 families were displaced from their homes by the dam, their houses disappearing beneath the lake that formed.
- The reservoir boosts the local tourist industry. Over 300,000 people visit each year. Activities include water skiing, yachting, forest walks, cycling, orienteering, camping and caravanning. Employment has been created in areas such as water sports, information centres and accommodation.
- The reservoir flooded a beautiful valley, an Area of Outstanding Natural Beauty.
- Schools use the reservoir for educational visits.
- Many jobs were created to build and maintain the reservoir.
- A variety of habitats are found around the reservoir, for example marshes and mudflats. Thousands of wildfowl come to Kielder Water over the winter. The area is home to ospreys again for the first time in over 200 years.
- North-east England now has a very reliable water supply.
- A hydroelectric power station, using the water released by Kielder reservoir, can generate 6 MW of power.
- Water can be sold to other parts of the UK during periods of shortage.

- Over 100,000 hectares of forest surrounds Kielder Water and is used to produce timber, employing 260 people.

- Over 1.5 million trees were lost when the valley was flooded.

- The water is not needed all the time and the reservoir has never been less than 90 per cent full.

- The road network between local villages was affected when the reservoir was created.

- Water releases are used to encourage salmon and sea trout to move up the river to spawn.

- The money spent on this scheme was so great it could have been better used to repair leaking pipes.

- There is a risk of contamination of one river ecosystem by transferring water from another.

- The forest at Kielder has been criticised for being too much of a monoculture (only one type of tree) – mainly Sitka Spruce.

## Sustainable management

The supply of water from the reservoir has to be sustainable. People today cannot deplete the water supply or damage the environment too much or the supply will not be adequate in the future. Although many people were initially sceptical, the reservoir helps to ensure long-term sustainable supply of water to many areas of north-east England and may, in the future, be used to transfer water to other parts of the country.

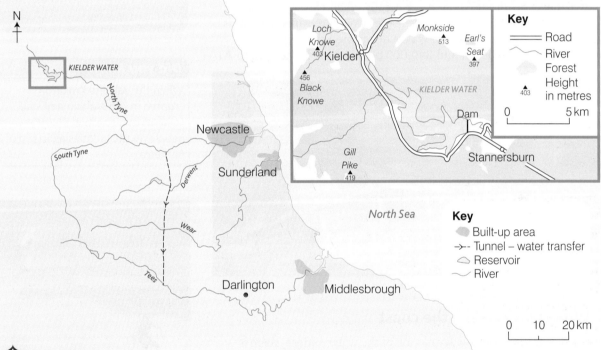

**E** Location of Kielder Water

### How waves form

Waves are usually formed by the wind blowing over the sea. Friction with the surface of the water causes ripples to form and these develop into waves. The stretch of open water over which the wind blows is called the **fetch**. The longer the fetch, the more powerful a wave can become.

Waves can also be formed more dramatically when earthquakes or volcanic eruptions shake the seabed. These waves are called tsunamis. In December 2004 giant tsunami waves devastated the countries bordering the Indian Ocean and 240,000 people were killed (photo **A**).

**In this section you will learn**

how waves are formed

why waves break at the coast

the characteristics of constructive and destructive waves.

**Did you know** ??????

When the volcano Krakatoa erupted in 1883, a tsunami said to be 35 m high killed 36,000 people on the Indonesian island of Sumatra.

**A** *The Indian Ocean tsunami 2004, Sri Lanka*

### When waves reach the coast

In the open sea, despite the wavy motion of the water surface, there is little horizontal transfer of water. It is only when the waves approach the shore that there is forward movement of water as waves break and wash up the **beach**. Diagram **B** shows what happens as waves approach the shore. Note how the seabed interrupts the circular orbital movement of the water. As the water becomes shallower, the circular motion becomes more elliptical. This causes the **crest** of the wave to rise up and then eventually to topple onto the beach. The water that rushes up the beach is called the **swash**. The water that flows back towards the sea is called the **backwash**.

**C** *Surfers on a beach in Newquay*

Top of wave moves faster

Wave begins to break

Water from previous wave returns

Water rushes up the beach

Circular orbit in open water

Friction with the seabed distorts the circular orbital motion

Increasingly elliptical orbit

Shelving seabed (beach)

**B** *Waves approaching the coast*

## Types of wave found at the coast

It is possible to identify two types of wave at the coast: **constructive waves** and **destructive waves**.

Constructive waves are waves that surge up the beach with a powerful swash. They carry large amounts of sediment and 'construct' the beach, making it more extensive. These are the waves that are loved by surfers (photo **C**). They are formed by distant storms, which can be hundreds of kilometres away. The waves are well spaced apart and are powerful when they reach the coast. Diagram **D** shows the typical characteristics of a constructive wave.

**Destructive waves** are formed by local storms close to the coast. They are so named because they 'destroy' the beach. Destructive waves are closely spaced and often interfere with each other, producing a chaotic, swirling mass of water. They rear up to form towering waves before crashing down onto the beach (diagram **E**). There is little forward motion (swash) when a destructive wave breaks, but a powerful backwash. This explains the removal of sediment and the destruction of the beach.

Constructive wave (smaller in height)  Crest  Strong swash: pushes material up the beach
Weak backwash: little erosion

**D** Constructive waves

Destructive wave (larger in height)  Weak swash: little beach-building
Strong backwash; scours the beach, pulling sand and shingle down the beach

**E** Destructive waves

### Key terms

**Fetch:** the distance of open water over which the wind can blow.

**Beach:** a deposit of sand or shingle at the coast, often found at the head of a bay.

**Crest:** the top of a wave.

**Swash:** the forward movement of a wave up a beach.

**Backwash:** the backward movement of water down a beach when a wave has broken.

**Constructive wave:** a powerful wave with a strong swash that surges up a beach.

**Destructive wave:** a wave formed by a local storm that crashes down onto a beach and has a powerful backwash.

### Activities

**1** Study photo **A**.

a What evidence is there in the photo that tsunamis are extremely powerful?

b Are there any safe places in the photo?

c Many of the people who died in 2004 lived in small coastal communities around the Indian Ocean. Why do people choose to live close to the sea?

**2** Study diagram **B**. Make a copy of it.

a Draw an arrow to show the waves approaching the coast.

b Write the labels 'swash' and 'backwash' in the correct places.

c What causes the waves to rise up and break on the beach?

d When waves break on a sandy or pebbly beach, the amount of backwash is often less than the amount of swash. Why do you think this is?

e On a pebble beach, larger pebbles are often found near the top of the beach with smaller ones near the bottom. Use your answer to d to suggest why this happens.

**3** Study photo **C**, which shows a constructive wave. They are sometimes known as a surging waves because they 'surge' up the beach.

a Why do surfers prefer constructive waves to destructive waves?

b Are constructive waves generated by storms close to the coast or a long way off?

c Wave power is being considered as a source of renewable energy off the north Cornwall coast. Do you think this is a good idea? Explain your answer.

### Study tip

Practise drawing a labelled sketch of a wave type. It is a good way to learn its features.

### ⚬⚬ links

You can find out more about the Indian Ocean tsunami at **www.guardian.co.uk/world/tsunami2004**.

Watch the surfers in action by webcam at **www.fistralsurfcam.com**.

# 4.2　Which land processes shape our coastline?

## ◼ Weathering

Processes of weathering affect rocks exposed at the coast. Rocks are weakened by elements of the weather such as rainfall and changes in temperature. There are three types of weathering:

1　Mechanical weathering, which involves the breakdown of rocks by physical forces such as heat, water and ice.

2　Chemical weathering, which involves the effect of chemicals in the breakdown of rocks. Rainwater, being slightly acidic, can dissolve certain rocks and minerals.

3　Biological weathering, which involves the action of plant roots and animals in helping to dislodge and breakdown rocks.

At the coast, freeze–thaw weathering is particularly effective if the rock exposed is porous (contains holes) and permeable (allows water to pass through it).

### Freeze–thaw weathering

**Freeze–thaw weathering**, or frost shattering as it is sometimes known, involves the action of water as it freezes and thaws in a crack or hole in the rock. It is a common process and operates wherever there is plenty of water and where temperatures fluctuate repeatedly above and below freezing point.

Look at diagram **A**. The process of freeze–thaw starts with liquid water collecting in cracks or holes (**pores**) in a rock. At night, this water freezes and expands by approximately 9 per cent. If the water is in a confined space, the expansion creates stresses within the rock, widening any cracks that already exist. When the temperature rises and the ice thaws, the water seeps deeper into the rock along newly formed cracks. After repeated cycles of freezing and thawing, fragments of rock may become detached and fall to the foot of the cliff. Look at photo **B**. The autumn of 2000 was exceptionally wet and the chalk rock became saturated with water. During late winter, long periods of frost weakened the rock, leading to several dramatic **rockfalls** along the south coast of England, including the one shown.

Rocks made up of calcium carbonate such as limestone and chalk, can be weathered chemically by a process of carbonation. Rainwater reacts with calcium carbonate, which then dissolves and is washed away in solution. This weakens the rock and causes it to break down.

## ◼ Mass movement

The rockfall shown in photo **B** is one example of mass movement. Mass movement is the downhill movement of material under the influence of gravity. In 1993, 60 m of **cliff** slid onto the beach near Scarborough in North Yorkshire, taking with it part of the Holbeck Hall Hotel (photo **C**). Teetering on the brink, the hotel had to be demolished. Diagram **D** describes some of the common types of mass movement found at the coast.

In this section you will learn

how weathering affects rocks at the coast

how processes of mass movement operate at the coast.

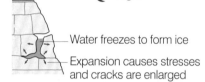

☀ Day

Coastal cliff

Water collects in cracks in rock

★ ☾ Night

Water freezes to form ice

Expansion causes stresses and cracks are enlarged

☀/★☾ Repeated freezing and thawing

Rock fragment breaks off and collects at the foot of the cliff

**A**　*The process of freeze–thaw weathering*

**B**　*Rockfall at Beachy Head, Sussex*

**C**　*Landslip at Holbeck Hall, Scarborough*

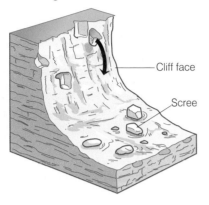

a **Rockfall** – fragments of rock break away from the cliff face, often due to freeze–thaw weathering

— Cliff face

Scree

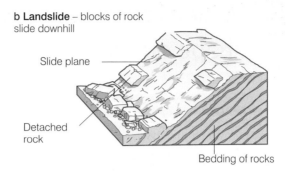

b **Landslide** – blocks of rock slide downhill

Slide plane

Detached rock

Bedding of rocks

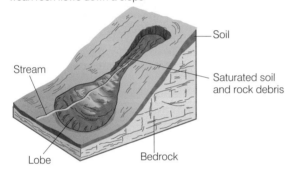

c **Mudflow** – saturated soil and weak rock flows down a slope

Soil

Stream

Saturated soil and rock debris

Lobe

Bedrock

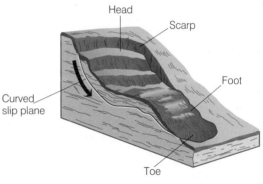

d **Rotational slip** – slump of saturated soil and weak rock along a curved surface

Head

Scarp

Foot

Curved slip plane

Toe

**D**   *Types of mass movement at the coast*

Both mass movement and weathering provide an input of material to the coastal system. Much of this material is carried away by the waves to be deposited elsewhere along the coast.

## Activity

Study photo **B**.

a   What do you think has happened recently to this stretch of coastline?

b   The chalk cliffs are affected by freeze-thaw weathering. How does freeze-thaw weathering operate?

c   What other weathering processes affect chalk? (Hint: chalk is made of calcium carbonate.) Describe how these processes work.

d   How might rockfalls such as that shown in the photo be a hazard to people?

e   What do you think will happen to the pile of rocks at the base of the cliff in the photo?

f   Imagine the local council has decided to place an information board at the top of the cliff to warn people of the dangers of cliff collapse. It wants to inform people why the cliff is dangerous. Design an information board explaining why the cliff is vulnerable to rockfalls. Use diagrams to illustrate your board and do not forget to warn people to keep well away from the cliff edge!

### Key terms

**Freeze-thaw weathering:** weathering involving repeated cycles of freezing and thawing.

**Pores:** holes in rock.

**Cliff:** a steep or vertical face of rock often found at the coast.

**Rockfall:** the collapse of a cliff face or the fall of individual rocks from a cliff.

### ∞ links

You can find out more about a Beachy Head rockfall in 1999 at **www.bbc.co.uk**. Enter 'Beachy Head rockfall' into the search box.

There is a spectacular video of a rockfall in Cornwall at **www.guardian.co.uk** search for 'massive Cornwall rockfall'.

# 4.3 | Which marine processes shape our coastline?

## Coastal erosion

When a wave crashes down on a beach or smashes against a cliff, it carries out the process of erosion (photo **A**). There are several processes of coastal erosion:

- **Hydraulic power** – this involves the sheer power of the waves as they smash onto a cliff (photo **A**). Trapped air is blasted into holes and cracks in the rock, eventually causing the rock to break apart. The explosive force of trapped air operating in a crack is called cavitation.
- **Corrasion** – this involves fragments of rock being picked up and hurled by the sea at a cliff. The rocks act like erosive tools by scraping and gouging the rock.
- **Abrasion** – this is the 'sandpapering' effect of pebbles grinding over a rocky platform, often causing it to become smooth.
- **Solution** – some rocks are vulnerable to being dissolved by seawater. This is particularly true of limestone and chalk, which form cliffs in many parts of the UK.
- **Attrition** – this is where rock fragments carried by the sea knock against one another, causing them to become smaller and more rounded.

## Coastal transportation

Four main types of sediment transportation can be identified (diagram **B**). The size and quantity of sediment transported by the sea depends on the strength of the waves and tidal currents. During storms, quite large pebbles can be flung up on to seawalls and promenades where they can cause damage to buildings and cars.

> **In this section you will learn**
>
> the processes of coastal erosion, transportation and deposition.

> **Key terms**
>
> **Hydraulic power**: the sheer power of the waves.
>
> **Corrasion**: the effect of rocks being flung at the cliff by powerful waves.

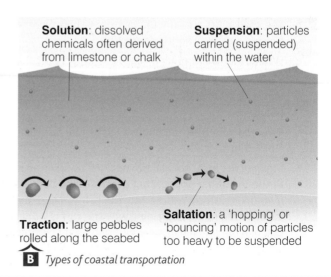

**Solution**: dissolved chemicals often derived from limestone or chalk

**Suspension**: particles carried (suspended) within the water

**Traction**: large pebbles rolled along the seabed

**Saltation**: a 'hopping' or 'bouncing' motion of particles too heavy to be suspended

**A** *Waves crashing onto cliffs in Iceland*

**B** *Types of coastal transportation*

The movement of sediment on a beach is largely determined by the direction of wave approach. Look at diagram **C**. Note that where the waves approach 'head on', sediment is moved up and down the beach. However, if the waves approach at an angle, sediment moves along the beach in a zig-zag pattern. This is called **longshore drift**.

**Key terms**

**Longshore drift**: the transport of sediment along a stretch of coastline caused by waves approaching the beach at an angle.

**Bay**: a broad coastal inlet often with a beach.

**Headland**: a point of usually high land jutting out into the sea.

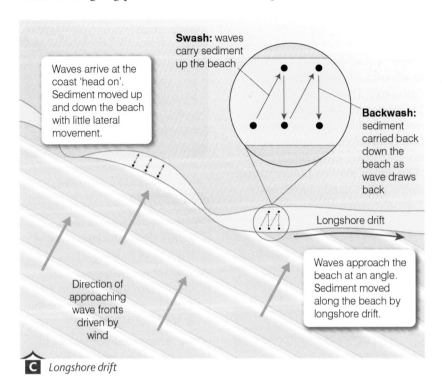

**Swash:** waves carry sediment up the beach

Waves arrive at the coast 'head on'. Sediment moved up and down the beach with little lateral movement.

**Backwash:** sediment carried back down the beach as wave draws back

Longshore drift

Waves approach the beach at an angle. Sediment moved along the beach by longshore drift.

Direction of approaching wave fronts driven by wind

**C**  *Longshore drift*

## ■ Coastal deposition

Coastal deposition takes place in areas where the flow of water slows down. Sediment can no longer be carried or rolled along and has to be deposited. Coastal deposition most commonly occurs in **bays**, where the energy of the waves is reduced on entering the bay. This explains the presence of beaches in bays and accounts for the lack of beaches at **headlands**, where wave energy is much greater.

*Study tip*

It is important to use the correct terms when describing coastal processes and to make sure that you can say what each one involves. Diagrams work well here.

**Activities**

1   Study diagram **B**. Draw an annotated diagram similar to this diagram to describe the processes of coastal erosion. Draw a simple diagram of a wave breaking against the foot of a cliff. Add detailed labels (annotations) in their correct places to describe the five processes of erosion.

2   Study diagram **C**.

a   What is meant by the term 'longshore drift'?

b   Why does longshore drift occur on some beaches but not on others?

c   Draw your own diagram or a series of diagrams to show how the process of longshore drift operates. Add labels to describe what is happening.

d   Imagine that you are going to carry out a fieldwork investigation to look for evidence of longshore drift along a stretch of coastline. What evidence would you look for? Explain your reasons.

## ⬯links

A good summary of coastal processes and landforms can be found at **www.georesources.co.uk/leld.htm**.

A fun animation using a rubber duck to show the process of longshore drift can be found at **www.geography-site.co.uk/pages/physical/coastal/longshore.html**.

# 4.4 What are the distinctive landforms resulting from erosion?

## ■ Headlands and bays

Cliffs rarely erode at an even pace. Sections of cliff that are particularly resistant to erosion stick out to form **headlands**. Weaker sections of coastline that are more easily eroded form **bays**. Diagram **A** shows the changes that take place to a coastline where rocks of different resistances meet the coast. Headlands are most vulnerable to the power of the waves, which explains the presence of erosional features such as cliffs and **wave-cut platforms.** In contrast, bays are often much more sheltered from the full fury of the sea. The waves are less powerful and deposition tends to dominate. This explains why a sandy beach is the most common feature found in bays.

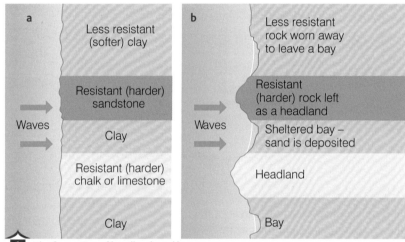

**A**  *The formation of headlands and bays*

### In this section you will learn

the formation and characteristics of features of coastal erosion (headlands and bays; wave-cut platforms and caves; caves, arches and stacks).

### Key terms

**Wave-cut platform**: a wide, gently sloping rocky surface at the foot of a cliff.

**Wave-cut notch**: a small indentation (or notch) cut into a cliff by coastal erosion roughly at the level of high tide.

**Cave**: a hollowed-out feature at the base of an eroding cliff.

**Arch**: a headland that has been partly broken through by the sea to form a thin-roofed arch.

**Stack**: an isolated pinnacle of rock sticking out of the sea.

## ■ Cliffs and wave-cut platforms

When waves break against a cliff, erosion close to the high-tide line takes a 'bite' out of the cliff to form a feature called a **wave-cut notch**. Over a long period of time – usually hundreds of years – the notch gets deeper until the overlying cliff can no longer support its own weight and it collapses. Through a continual sequence of wave-cut notch formation and cliff collapse, the cliff line gradually retreats. In its place will be a gently sloping rocky platform called a wave-cut platform (photo **B**).

A wave-cut platform is typically quite smooth due to the process of abrasion, but in some places it may be pockmarked with rock pools. During long periods of constructive waves, the wave-cut platform may become covered by sand or shingle. Destructive waves associated with local winter storms remove the beach once again, exposing the wave-cut platform.

**B**  *Wave-cut platform and cliff near Beachy Head*

## ■ Caves, arches and stacks

Lines of weakness in a headland, such as joints or faults, are particularly vulnerable to erosion. The energy of the waves gouges out the rock along a line of weakness to form a **cave** (diagram **C**). Over time, erosion may lead to two back-to-back caves breaking through a headland to form an **arch**. Gradually, the arch is enlarged by erosion at the base and sides and by weathering processes acting on the roof. The roof collapses eventually to form an isolated pillar of rock known as a **stack**.

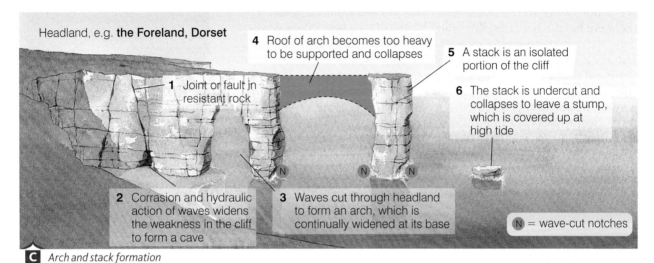

Headland, e.g. **the Foreland, Dorset**

**4** Roof of arch becomes too heavy to be supported and collapses

**5** A stack is an isolated portion of the cliff

**1** Joint or fault in resistant rock

**6** The stack is undercut and collapses to leave a stump, which is covered up at high tide

**2** Corrasion and hydraulic action of waves widens the weakness in the cliff to form a cave

**3** Waves cut through headland to form an arch, which is continually widened at its base

N = wave-cut notches

**C**  *Arch and stack formation*

Study photo **B**.

a  Draw a simple sketch of the cliff and wave-cut platform in the photo.

b  Add labels to identify the cliff and the wave-cut platform. Suggest where you think the wave-cut notch is located at the foot of the cliff.

c  Add labels to indicate where the processes of erosion operate at high tide. The rock in the photo is chalk.

d  Use a series of simple annotated diagrams to show how a cliff (such as the one in the photo) is undercut by the sea and then collapses to form a wave-cut platform. Include reference to the processes and landforms in your annotations.

e  Under what conditions might a wave-cut platform become covered by sand or shingle?

f  Is it a good time to go rock-pooling after a period of stormy weather? Explain your answer.

∞**links**

Good diagrams of coastal features can be found at **www.georesources.co.uk/leld.htm**.

Check out Geography at the Movies at **www.gatm.org.uk**.

# 4.5 What are the distinctive landforms resulting from deposition?

## Beaches

Beaches are accumulations of sand and shingle (pebbles) found where deposition occurs at the coast. Sandy beaches are often found in sheltered bays, where they are called bay head beaches. When waves enter these bays, they tend to bend to mirror the shape of the coast. This is called wave refraction (photo **A**). The way the water gets shallower as the waves enter the bays causes this to happen. Wave refraction spreads out and reduces the wave energy in a bay, which is why deposition occurs here.

> **In this section you will learn**
>
> the formation and characteristics of features of coastal deposition (beaches, spits and bars).

> **Key terms**
>
> **Spit**: a finger of new land made of sand or shingle, jutting out into the sea from the coast.
>
> **Salt marsh**: low-lying coastal wetland mostly extending between high and low tide.
>
> **Bar**: a spit that has grown across a bay.

Sea

Bending (refraction) of waves

Land

**A** *Wave refraction at the head of a bay*

Elsewhere along the coast, pebble beaches may form. These are most commonly found in areas where cliffs are being eroded and where there are higher-energy waves, such as along the south coast of the UK.

Ridges or berms are common characteristics of a beach (photo **B**). They are small ridges coinciding with high-tide lines and storm tides. Some beaches may have several berms, each one representing a different high-tide level.

## Spits

A **spit** is a long, narrow finger of sand or shingle jutting out into the sea from the land (diagram **C**). Spits are common features across the world. As sediment is transported along the coast by longshore drift, it becomes deposited at a point where the coastline changes direction or where a river mouth occurs. Gradually, as more and more sediment is deposited, the feature extends into the sea. Away from the coast, the tip is affected by waves approaching from different directions and the spit often becomes curved as a result.

**B** *Berms on a beach at Deal on the Kent coast*

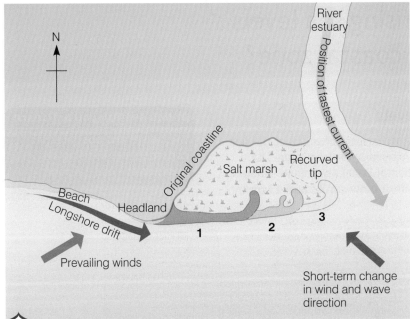

N

River estuary

Position of fastest current

Original coastline

Salt marsh

Recurved tip

Beach

Headland

Longshore drift

Prevailing winds

1    2    3

Short-term change in wind and wave direction

**C**   *The formation of a spit*

**D**   *Coastal bar at Slapton Ley, Devon*

Over time, the sediment breaks the surface to form new land and a spit is formed. It soon becomes colonised by grass and bushes, and eventually trees will grow. On the landward sheltered side of a spit where the water is a calm, mudflats and **salt marshes** form. These are important habitats for plants and birds. Being close to sea level, spits are vulnerable to erosion, especially during storms.

### Bars

Occasionally, longshore drift may cause a spit to grow right across a bay, trapping a freshwater lake or lagoon behind it. This feature is called a **bar** (photo **D**). In the UK some offshore bars have been driven onshore by rising sea levels following ice melt at the end of the last glacial period some 10,000 years ago. This type of feature is called a barrier beach. Chesil Beach in Dorset is one of the best examples of this feature.

**Study tip**

A spit is a land feature, so it does not get covered at high tide. On a map, it will be clearly bordered by the high tide line. Be careful not to confuse it with sediment exposed only at low tide.

**links**

Information on Dungeness spit in Washington State in the USA, can be found at **www.reefnews.com/reefnews/oceangeo/washngtn/dnwr.html.**

**Activities**

1   Study photo **A**.
a   What is wave refraction? Draw a simple sketch based on the photo to support your answer.
b   How does wave refraction explain the formation of a bay head beach?
c   A sandy beach is more likely to be found in a sheltered bay, whereas a shingle (pebble) beach is more likely to be found on an open and exposed stretch of coastline. Explain why.

2   Study photo **B**. What are berms and how do they form on a beach? Use a simple diagram to help your explanation.

3   Study diagram **C** and photo **D**.
a   What is a spit?
b   Draw a series of three simple diagrams to show the formation of a spit. Add detailed labels to describe what is happening using the text and the diagram to help you.
c   What is the difference between a spit and a bar?
d   What is the difference between a bar and a barrier beach?

## 4.6 How will rising sea levels affect the coastal zone?

### ■ The causes of rising sea levels

One of the effects of global warming is sea-level change. Over the last 15 years, global average sea levels have risen by 3 mm a year. The latest estimates from the Intergovernmental Panel on Climate Change (IPCC) suggest a rise in global sea levels of between 28 and 43 cm by the end of the century.

> **In this section you will learn**
>
> the causes and possible consequences of rising sea levels on the coastal zone.

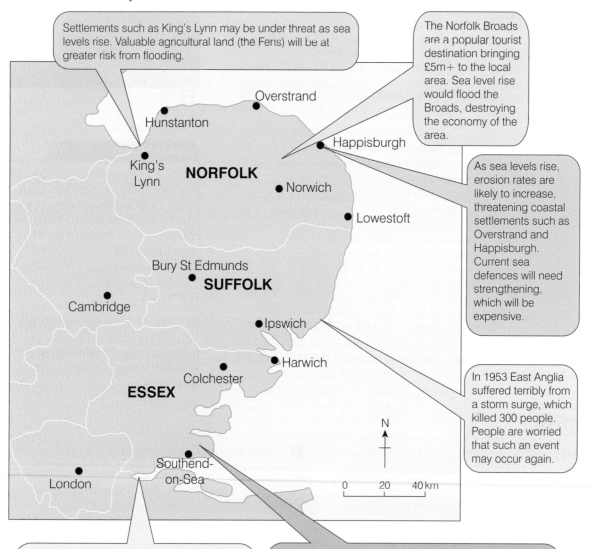

Settlements such as King's Lynn may be under threat as sea levels rise. Valuable agricultural land (the Fens) will be at greater risk from flooding.

The Norfolk Broads are a popular tourist destination bringing £5m+ to the local area. Sea level rise would flood the Broads, destroying the economy of the area.

As sea levels rise, erosion rates are likely to increase, threatening coastal settlements such as Overstrand and Happisburgh. Current sea defences will need strengthening, which will be expensive.

In 1953 East Anglia suffered terribly from a storm surge, which killed 300 people. People are worried that such an event may occur again.

The Thames Barrier currently protects buildings worth £80bn. It will probably need to be replaced in the next 30 to 50 years. As sea levels rise, large areas of the lower Thames estuary will be at risk from flooding, affecting housing, industry and farmland.

Low-lying mudflats and marshes in Essex are particularly vulnerable to sea-level rise. Areas of salt marsh are being squeezed between sea walls and rising sea. Some 22 per cent of East Anglia's salt marsh could be lost by 2050. In some places, managed retreat (a controversial political decision) will breach sea walls to allow deliberate flooding so that salt marshes can reform.

**A** *Possible impact of sea-level rise in East Anglia*

The main cause of sea-level rise is thermal expansion of the seawater as it absorbs more heat from the atmosphere. The melting of ice on the land, for example from glaciers on Greenland, will increase the amount of water in the oceans but scientists do not believe that this will significantly affect sea levels. Melting sea ice, such as from the Arctic, will have no direct effect on sea levels.

The actual amount of sea-level rise will vary from place to place. It is complicated by relative rises and falls in the level of the land as well as variations in the amount of deposition occurring at the coast. The amount of sea-level rise will also depend on the rate of global warming. However uncertain the future may be, most scientists agree that sea levels will rise and that coastal zones will be at an increasing risk from wave attack and flooding.

In the UK, East Anglia (diagram **A**) is likely to be hardest hit as rising sea levels threaten coastal defences and natural ecosystems. Elsewhere in the world, vast areas of low-lying coastal plains such as Bangladesh and whole chains of islands such as the Maldives and Tuvalu could disappear.

Already, two Pacific islands have been submerged, and many others have been flooded. Since more than 70 per cent of the world's population live on coastal plains, the effects of rising sea levels are likely to be devastating.

> **Did you know** ??????
>
> If sea levels rise by 1m, several major cities of the world will be affected by flooding including London, Tokyo and New York.

> **Study tip**
>
> Be sure to understand the meaning of economic (money), social (people), environmental (natural world) and political (decision-making) when describing the possible impacts of sea-level rise.

## Activities

**1** Study diagram **A**.

a Describe some of the likely economic impacts of sea-level rise in East Anglia.

b Suggest some likely social and political (decision-making, often by the government) impacts of sea-level rise. Some of these will be linked to the economic factors identified in a.

c With the aid of a diagram, describe how salt marshes could be lost as sea levels rise.

d Why is there concern about loss of salt marsh environments in East Anglia?

**2** This activity will involve internet research.

a Select a coastal zone under threat from sea-level rise, such as Bangladesh, the Mississippi delta in the USA or the Maldives.

b Conduct an internet search ('sea level rise + Maldives', for example) to find out about the economic, social, environmental and political impacts of sea-level rise on your chosen locality.

c Draw a diagram similar to diagram **A**, using a photo or map of your locality in the centre.

**B**   *The islands of the Maldives are under threat from sea-level rise*

## ∞ links

http://thelondonflood.com is an excellent source of information. www.floodlondon.com looks at what might happen if London was flooded.

# 4.7 How can cliff collapse cause problems for people at the coast?

Various factors can contribute to cliff collapse. These include weathering processes such as heavy rainfall that can saturate the land and make it unstable, mass movement such as **sliding** and **slumping**, which is more likely if the land is made of soft weak rock types, and of course the power of the waves continually crashing against the cliffs and undercutting them from below. Various stretches of UK coastline are vulnerable to cliff collapse, including Christchurch Bay in Hampshire, where Barton-on-Sea is sited.

> **In this section you will learn**
>
> the causes and consequences of cliff collapse for people living in the coastal zone.

## Case study

### Barton-on-Sea, Hampshire

Locate the small settlement of Barton-on-Sea on map **B**. This stretch of coastline in Christchurch Bay has long been affected by coastal erosion and cliff collapse. Over the years a number of buildings, and most recently a café, have been lost to the sea.

Extensive coastal defences have been built to try to prevent coastal erosion. However, in 2008 a fresh **landslip** occurred (photo **A**). This has once again raised concerns among local residents about the vulnerability of this part of the coast to cliff collapse. An older development of houses in Barton Court is just 20 m from the cliff edge. The local authority predicts that the houses will be lost in the next 10 to 20 years.

The cliffs at Barton-on-Sea are prone to collapse due to a number of factors:

- The rocks are weak sands and clays. They are easily eroded by the sea and have little strength to resist collapse.

- The arrangement of the rocks (permeable sands on top of impermeable clay) causes water to 'pond-up' within the cliffs. This increases the weight of the cliffs. The increase in water pressure within the cliffs (called pore water pressure) encourages collapse.

**A** *Cliff collapse at Barton-on-Sea, 2008*

- This stretch of coastline faces the direct force of the prevailing south-westerly winds. With a long fetch, the waves approaching Barton-on-Sea are powerful and can carry out a great deal of erosion. Rates of erosion have been as much as 2 m a year in places.

- Several small streams (with the local name 'Bunny') flow towards the coast but disappear into the permeable sands before they reach the sea. This adds to the amount of water in the cliffs.

- Buildings on the cliff top have increased the weight on the cliffs, making them more vulnerable to collapse. They can also interfere with drainage.

## ⌒⌒ links

A website with fantastic photos is www.southampton.ac.uk/~imw/barteros.htm.

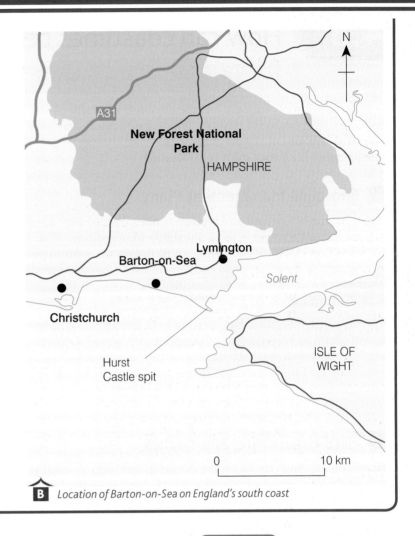

**B**   *Location of Barton-on-Sea on England's south coast*

## Activity

Study photo **A**.

a  What evidence is there that there has been a cliff collapse?

b  Estimate the drop in the grass surface due to the collapse.

c  Look closely at the rock material forming the cliffs. How strong do you think it is? Explain your answer.

d  What action has been taken to keep people at a safe distance from the cliff?

e  Using evidence from the photo, suggest how this and future cliff collapses might affect people living in the area.

f  How does the cliff collapse affect the local environment?

g  Draw a simplified sketch of the cliffs at Barton-on-Sea and add annotations to identify the factors that have led to the cliff collapse.

## Key terms

**Sliding**: a type of mass movement involving material moving downhill on a flat surface (a landslide).

**Slumping**: a type of mass movement involving material moving downhill under its own weight.

**Landslip**: a type of mass movement common at the coast involving material slipping downhill usually along a curved slip surface.

## Study tip

Cliff collapse usually results from the interaction of several factors. Make this clear in any explanation you are required to give.

# 4.8 How can coastlines be managed?

The coastal zone needs to be managed in order to maintain a balance between the forces of nature and the demands of people. The coastline is under threat from cliff collapse, flooding and future sea-level rise. With millions of people living in the coastal zone, sustainable management is an important consideration.

## Shoreline Management Plans

The coastline of England and Wales has been divided into a number of self-contained sediment cells. A **Shoreline Management Plan (SMP)** has been developed for each area, which details the natural processes, environmental considerations and human uses. Coastlines at risk from erosion or flooding have been identified and plans put in place to cope with the issues.

In most cases, a decision has been made to 'hold the line'. This means taking action to keep the line of the coast as it is now. Occasionally, planners decide to 'advance the line' of the coast to afford greater protection by, for example, increasing the size of the beach. In taking action, authorities have the option of hard or soft engineering approaches.

## Hard engineering approaches

**Hard engineering** involves using artificial structures to control the forces of nature. For many decades people have used hard engineering structures such as sea walls (photo **A**) and groynes (photo **B**) to try to control the actions of the sea and protect property from flooding and erosion. Table **C** outlines the common hard engineering approaches used in coastal management.

> ### In this section you will learn
>
> the options of coastal management including hard and soft engineering and managed retreat.

> ### Key terms
>
> **Shoreline Management Plan (SMP)**: an integrated coastal management plan for a stretch of coastline in England and Wales.
>
> **Hard engineering**: building artificial structures such as sea walls aimed at controlling natural processes.
>
> **Soft engineering**: a sustainable approach to managing the coast without using artificial structures.

> ### Did you know ???????
>
> In 1902 a sea wall was constructed at Galveston, USA following the devastating hurricane of 1900. The wall is 16 km long, 5.2 m high and 4.9 m thick at its base. So far, it has never been overtopped.

**A** *The sea wall at Dawlish, Devon*

**B** *Groynes at Eastbourne*

**C** *Hard engineering schemes*

| Hard engineering | Description | Cost | Advantages | Disadvantages |
|---|---|---|---|---|
| **Sea wall** | Concrete or rock barrier built at the foot of cliffs or at the top of a beach. Has a curved face to reflect the waves back into the sea, usually 3–5m high. | Up to £10 million per km (south sea zones). | • Effective at stopping the sea.<br>• Often has a walkway or promenade for people to walk along. | • Can be obtrusive and unnatural to look at.<br>• Very expensive and has high maintenance costs. |
| **Groynes**<br>Beach sand | Timber or rock structures built out to sea from the coast. They trap sediment being moved by longshore drift, and broaden the beach. The wider beach acts as a buffer to the incoming waves, reducing wave attack at the coast. | Up to £5,000 per metre. | • Result in a bigger beach, which can enhance the tourist potential of the coast.<br>• Provide useful structures for people interested in fishing.<br>• Not too expensive. | • In interrupting longshore drift, they starve beaches downdrift, often leading to increased rates of erosion elsewhere. The problem is not so much solved as shifted.<br>• Groynes are unnatural and rock groynes in particular can be unattractive. |
| **Rock armour** | Piles of large boulders dumped at the foot of a cliff. The rocks force waves to break, absorbing their energy and protecting the cliffs. The rocks are usually brought in by barge to the coast. | Approximately £1,000–£4,000 per metre. | • Relatively cheap and easy to maintain.<br>• Can provide interest to the coast. Often used for fishing. | • Rocks are usually from other parts of the coastline or even from abroad. Can be expensive to transport.<br>• They do not fit in with the local geology.<br>• Can be very obtrusive. |

Hard engineering approaches are less commonly used today. Not only are they expensive and involve high maintenance costs, they are also obtrusive and unnatural. They tend to interfere with natural coastal processes and can cause destructive knock-on effects elsewhere. In altering wave patterns, for example, erosion can become concentrated further down the coast leading to new problems of cliff collapse.

## ■ Soft engineering approaches

**Soft engineering** approaches (table **D**) try to fit in and work with the natural coastal processes. They do not involve large artificial structures. Often more 'low key' and with low maintenance costs – both economically and environmentally – soft engineering approaches such as beach nourishment (photo **E**) are more sustainable. They are usually the preferred option of coastal management.

> **Key terms**
>
> **Sea wall**: concrete or rock barrier built at the foot of cliffs or at the top of a beach.
>
> **Groyne**: timber or rock structure built out to sea to trap sediment being moved by longshore drift.
>
> **Rock armour**: piles of large boulders dumped at the foot of a cliff to protect it by forcing waves to break and absorbing their energy.

D  *Soft engineering schemes*

| Soft engineering | Description | Cost | Advantages | Disadvantages |
|---|---|---|---|---|
| Beach nourishment | The addition of sand or shingle to an existing beach to make it higher or broader. The sediment is usually obtained locally so that it blends in with the existing beach material. Usually brought onshore by barge. | Approximately £3,000 per metre | • Relatively cheap and easy to maintain.<br>• Blends in with existing beach.<br>• Increases tourist potential by creating a bigger beach. | • Needs constant maintenance unless structures are built to retain the beach. |
| Dune regeneration | Sand dunes are effective buffers to the sea yet they are easily damaged and destroyed, especially by trampling. Marram grass can be planted to stabilise the dunes and help them to develop. Areas can be fenced to keep people off newly planted dunes. | Approximately £2,000 per 100 m | • Maintains a natural coastal environment that is popular with people and wildlife.<br>• Relatively cheap. | • Time-consuming to plant the marram grass and fence off areas.<br>• People do not always respond well to being prohibited from accessing certain areas.<br>• Can be damaged by storms. |
| Marsh creation (managed retreat) | This involves allowing low-lying coastal areas to be flooded by the sea to become salt marshes. This is an example of managed retreat. Salt marshes are effective barriers to the sea. | Depends on the value of the land. Arable land costs somewhere in the region of £5,000 to £10,000 per hectare | • A cheap option compared with maintaining expensive sea defences that might be protecting relatively low-value land.<br>• Creates a much-needed habitat for wildlife. | • Land will be lost as it is flooded by sea water.<br>• Farmers or landowners will need to be compensated. |

## ■ Managed retreat

A further option for coastal management is to allow for some retreat of the coastline. This is called **managed retreat** or coastal realignment. It is a real option if there is a high risk of flooding or cliff collapse and where the land is relatively low value. Poor quality grazing land, for example, is seldom worth protecting if the costs of defences outweigh the benefits of protection (see Marsh creation in table **D**). This appraisal is known as a cost–benefit analysis. With sea levels forecast to rise, this approach is likely to become an increasingly popular option.

E  *Beach nourishment in operation at Poole in Dorset*

> **Key terms**
>
> **Managed retreat**: allowing controlled flooding of low-lying coastal areas or cliff collapse in areas where the value of the land is low.

> **Study tip**
>
> Be clear about the differences between hard engineering and soft engineering. Appreciate that there is a great deal of debate about the option of managed retreat. Make sure that you know both sides of the argument.

## Activity

Use photos **A**, **B** and **E**, and tables **C** and **D**, to help you answer the following questions.

a Why is a sea wall an example of hard engineering?

b What is the purpose of a sea wall?

c What are the advantages and disadvantages of a sea wall?

d What are groynes and what is their purpose? Draw a simple diagram to support your answer.

e How effective do you think the groynes are in photo **B**?

f Beach nourishment is a good example of soft engineering. What is soft engineering?

g With reference to photo **E**, outline some of the advantages of beach nourishment.

h Imagine a local council wishes to defend a 1 km stretch of coastline. Use tables **C** and **D** to calculate comparative costs for a sea wall, groynes, rock armour and beach nourishment.

i Why is the economic cost only one of several considerations when deciding which coastal defence measure to adopt?

# Coastal defences at Minehead

Minehead on the north coast of Somerset is one of the region's premier tourist resorts. It is home to a large Butlin's resort and every year it is visited by many thousands of tourists.

By the early 1990s it became clear that the current sea defences were going to be inadequate in the future. Storm damage was estimated to be £21m if nothing were done. The Environment Agency developed a plan to defend the town and improve the amenity value. Work started in 1997 and the sea defences were officially opened in 2001. The total cost was £12.3m, which represents a considerable saving on the potential losses due to storm damage.

The main features of the scheme (diagram **F**) are:

- A 0.6m high sea wall with a curved front to deflect the waves. It has a curved top to deter people from walking on it and its landward side is faced with attractive local red sandstone.
- Rock armour at the base of the wall to dissipate some of the wave energy.
- Beach nourishment (sand) to build up the beach by 2 m in height. This forces the waves to break further out to sea and provides an excellent sandy beach for tourists.
- Four rock groynes to help retain the beach and stop longshore drift moving sand to the east.
- A wide walkway with seating areas alongside the sea wall. This is popular with tourists and local people.

The scheme has been extremely successful. Not only does it protect the town from storms and high tides, but it has also enhanced the seafront by creating an attractive beach environment (photo **G**).

Case study

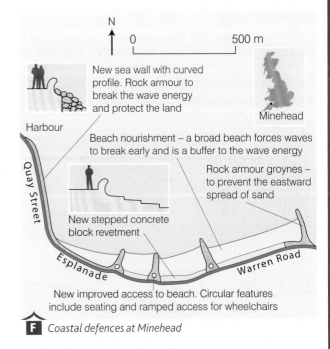

**F** *Coastal defences at Minehead*

**G** *The beach at Minehead*

# 4.9   What are salt marshes and why are they special?

Salt marshes are areas of periodically flooded low-lying coastal wetlands. They are often rich in plants, birds and animals.

A salt marsh begins life as an accumulation of mud and silt in a sheltered part of the coastline, for example in the lee of a spit or bar. As more deposition takes place, the mud begins to break the surface to form mudflats. Salt-tolerant plants such as cordgrass soon start to colonise the mudflats. These early colonisers are called **pioneer plants**. Cordgrass is tolerant of the saltwater and its long roots prevent it from being swept away by the waves and the tides. Its tangle of roots also helps to trap **sediment** and stabilise the mud.

As the level of the mud rises, it is less frequently covered by water. The conditions become less harsh as rainwater begins to wash out some of the salt and decomposing plant matter improves the fertility of the newly forming soil. New plant species such as sea asters start to colonise the area and gradually, over hundreds of years, a succession of plants develops. This is known as a **vegetation succession** (diagram **A**).

> ### In this section you will learn
>
> the characteristics of a salt marsh environment
>
> sustainable approaches to the management of salt marshes.

> ### Key terms
>
> **Pioneer plant**: the first plant species to colonise an area that is well adapted to living in a harsh environment.
>
> **Sediment**: loose rock debris that has been weathered or eroded before being transported and then deposited.
>
> **Vegetation succession**: a sequence of vegetation species colonising an environment.

> ### Did you know ?????
>
> In France, salt is produced commercially from salt marshes.

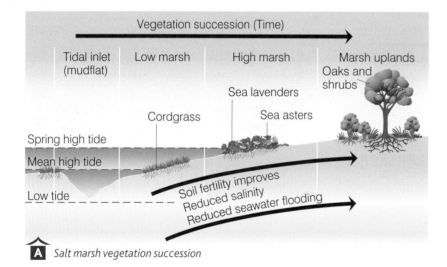

**A**  Salt marsh vegetation succession

## Keyhaven Marshes, Hampshire

*Case study*

Keyhaven Marshes is an area of salt marsh formed in the lee of Hurst Castle spit. It supports a range of habitats including grassland, scrub, salt marsh and reed beds. This variety of habitats accounts for a rich diversity of wildlife in the area (table **B**).

In common with many areas of salt marsh in the UK, Keyhaven Marshes is under threat:

■ The salt marsh is retreating by up to 6 m a year. Although the causes of this are not yet fully understood, further sea-level rise threatens a 'squeeze'

of the salt marsh as it lies between a low sea wall built in the early 1990s and the encroaching sea.

■ The salt marsh has been under threat from the breaching of Hurst Castle spit during severe storms. In December 1989 storms pushed part of the shingle ridge over the top of the salt marsh, exposing 50 to 80 m to the full fury of the sea. It was eroded in less than three months.

■ Increasing demands for leisure and tourism have meant that increasing numbers of people wish to visit the marshes. Careful management is required to

**B**  *Common wildlife species found at Keyhaven Marshes*

| Species detail | Image | Species detail | Image |
|---|---|---|---|
| Plant: cordgrass – spiky, untidy-looking grass that grows fast on mudflats |  | Bird: ringed plover – feeds intertidally and nests on the salt marsh |  |
| Plant: sea lavender – attractive, colourful flowers attract wildlife |  | Butterfly: common blue – resident butterfly commonly found on higher marshes |  |
| Bird: oystercatcher – feeds and nests in salt marshes |  | Spider: wold spider – clings for hours to submerged stems of cordgrass waiting for low tide and food |  |

prevent damage by trampling, parking and pollution. The area is popular with mariners who use the many creeks to moor their boats.

■ In 1996 rock armour and beach nourishment were used to increase the height and width of the spit in an attempt to stop breaching. Since the completion of the £5m sea defences, the spit has not been breached and Keyhaven Marshes seems safe … at least for the time being.

■ Keyhaven Marshes has been nationally recognised as an important site for wildfowl and wading birds. The area is officially a Site of Special Scientific Interest (SSSI) and part of the salt marsh is also a National Nature Reserve. This means that the area is carefully monitored and managed to maintain its rich biodiversity. Access is limited and development restricted.

■ For the future, with sea levels expected to rise by 6 mm a year, the big issue concerns the 'squeeze' between the low sea wall and the rising sea.

**C**  *The salt marsh at Keyhaven*

## ∞ links

An excellent case study based on Chichester harbour can be accessed at **www.conservancy.co.uk/learn/ wildlife/saltmarsh.htm**.

### Activity

**1** Study diagram **A**, table **B** and photo **C**.

a What is a salt marsh and what makes it a special habitat?

b Why do salt marshes often develop in the shelter of spits?

c Why is cordgrass well suited to be a pioneer species?

d How does cordgrass improve the salt marsh environment to enable other plant species to grow?

e What is meant by a vegetation succession?

f How do the environmental conditions of a salt marsh change as the vegetation succession proceeds?

g Suggest ways that people's actions might affect a natural salt marsh vegetation succession.

# 5 Contemporary population issues

## 5.1 How does population grow?

### Exponential growth

In 2011 the world's population topped 7 billion people. Since about 1900 the world's population has grown exponentially. This means that the rate of growth has become increasingly rapid. Between AD 1 and AD 1000 growth was slow, but in the last thousand years it has been dramatic. By 2000, there were 10 times as many people living as there had been 300 years before in 1700. Not only is population increasing, but the rate of increase is becoming greater.

Population grew especially quickly during the late 20th and early 21st centuries. Between 2008 and 2009, 220,980 people were added to world population *every day*. Growth is predicted to continue, but now the rate is slowing down. Population is likely to rise to 8.92 billion by 2050 and finally peak a century later in 2150 at 10.8 billion. This should be followed by a more stable period of **zero growth** or even **natural decrease (ND)**.

Population growth is usually shown as a line graph. **Exponential growth** produces a line that becomes steeper over time, taking the shape of a letter J. Today, growth rates are slowing down (although the numbers being added daily are still high), so the shape of the graph is levelling off into an S curve (graph **A**).

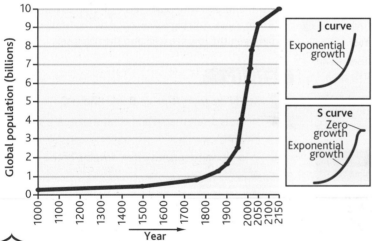

**A** *Global population growth*

**B** *European population, 1800–2010*

| Year | Population (millions) | Year | Population (millions) |
|------|----------------------|------|----------------------|
| 1800 | 203 | 1985 | 706 |
| 1900 | 408 | 1995 | 727 |
| 1950 | 547 | 2005 | 725 |
| 1975 | 676 | 2010 | 830 |

**Key terms**

**Zero growth**: a population in balance. Birth rate is equal to death rate, so there is no growth or decrease.

**Natural decrease (ND)**: the death rate minus the birth rate per 1000 per year.

**Exponential growth**: a pattern where the growth rate constantly increases – often shown as a J-curve graph.

**Activity**

1 Study table **B**.

a Use the data from the table to draw a growth curve for Europe. Plot the year along the x-axis and the population along the y-axis.

b Name the shape of the graph you have drawn. Why is this name used?

c Describe the population growth pattern that you have drawn. Give details of the graph's shape and quote figures from the graph to support your answer. Use terms such as *exponential*, *J-shaped, S-shaped, levelling off, zero growth* and *population decline*.

# 5.2  What is the demographic transition model?

Demography is the study of population. Transition simply means change. The **demographic transition model (DTM)** is shown in diagram **A**.

The model explains birth and death rate patterns across the world and through time. It includes the main period of a country's development and shows the links between demographic and economic changes. The diagram is divided into five stages, showing change from high birth and death rates in Stage 1 to much lower ones in Stages 4 and 5. Originally, the model was designed to explain population change in countries at further stages of development, but it has since been used to explain events in all countries. This allows us to compare different patterns of demographic and economic development.

**In this section you will learn**

about the demographic transition model and how it can be interpreted

about reasons for changes in world population.

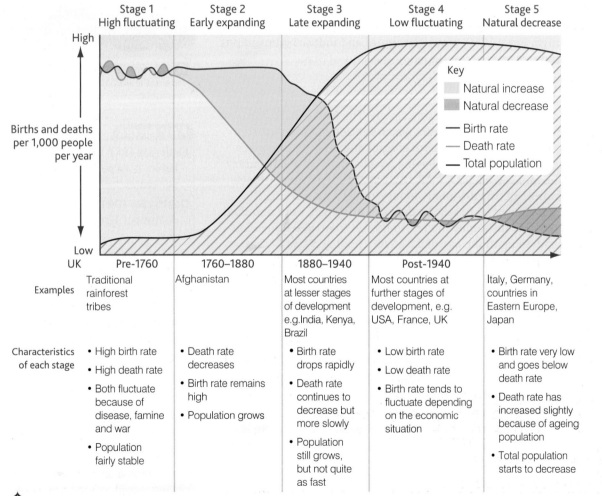

 *Demographic transition model*

**Key term**

**Demographic transition model (DTM):** a theoretical model that shows changes in population information (birth and death rates and population growth) over a period of time.

## Characteristics of each stage

### Stage 1 – high fluctuating

The high fluctuating stage occurs in societies where there is little medicine, low **life expectancy** and no means of birth control. Remote rainforest areas of Amazonia and Indonesia are the only locations where this stage might happen today. These are traditional societies, largely cut off from the rest of the world. The UK was at Stage 1 before around 1760.

### Stage 2 – early expanding

The key factor that indicates the change from Stage 1 to Stage 2 is a decrease in **death rate**. Improvements in medicine and hygiene cure some diseases and prevent others. Life expectancy increases. The gap between **birth rate** and death rate results in population growth. The farther apart the lines are on the graph, the greater the rate of growth. Most economies in Stage 2 are agricultural. Children are needed to work the land, as they can produce more food than they eat. This keeps birth rates high. From 1760, new agricultural and industrial inventions and medical discoveries in the UK led to the start of Stage 2.

### Stage 3 – late expanding

Death rate continues to fall, but more slowly. The key factor at the start of Stage 3 is a decrease in birth rate, which is often quite rapid. This is due to both the availability of birth control and economic changes, which mean people benefit from having smaller families. As a country develops, children become economic costs instead of economic assets. In other words, they cost the family money rather than earning it! When children have to go to school, they can no longer work and their education may cost the family money. In the UK, the 1870 Education Act made children up to the age of 12 go to school at a cost of one penny a week per child. The UK entered Stage 3 around 1880.

### Stage 4 – low fluctuating

Birth and death rates are both low. The lines on the graph are close to each other and birth rate varies according to the economic situation. When the economy is growing and people have jobs and earn a good living, they are more likely to afford children. In times of unemployment and low wages, people tend to postpone having a family until times are better. In the UK in the 1960s, the economy was growing and the birth rate rose slightly. But, in the 1970s, with world economic recession, the birth rate fell again. Overall, there is still population growth, but it is slow. Birth rates in the USA have been declining since a record high in 2007. It is thought that this decline could be explained by the economic recession that began in 2008.

### Stage 5 – natural decrease

Many Eastern and a few Western European countries are at Stage 5, but for different reasons. The UK remains at Stage 4. Death rate rises (as indeed it has in the UK too) because the population includes more elderly people. In Eastern Europe an uncertain economy discourages people from having babies, while Western European economies give

**B**  *The introduction of machines heralds the start of the agricultural revolution*

**Key terms**

**Birth rate (BR)**: the number of babies born per 1,000 people per year.

**Death rate (DR)**: the number of deaths per 1,000 people per year.

**Natural increase (NI)**: the birth rate minus the death rate per 1000 per year.

**Life expectancy**: the number of years a person is expected to live, usually taken from birth.

**Study tip**

Learn the characteristics of the five stages carefully and be able to identify the differences. Know examples of countries in each stage today and know when the UK was at each of Stages 1 to 4.

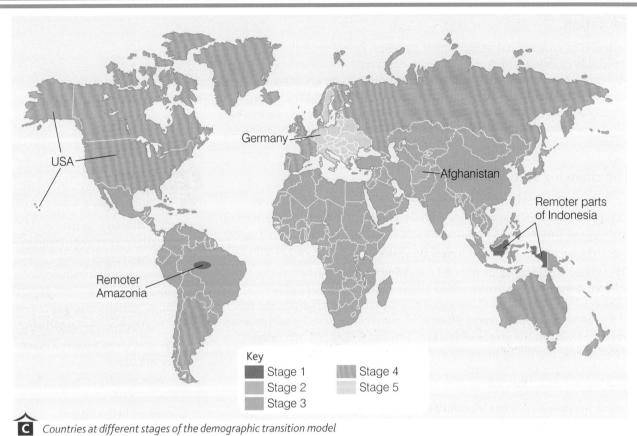

**Key**
- Stage 1
- Stage 2
- Stage 3
- Stage 4
- Stage 5

USA
Germany
Afghanistan
Remoter parts of Indonesia
Remoter Amazonia

**C**  *Countries at different stages of the demographic transition model*

young women so many career opportunities that they decide to be childless or to postpone motherhood. For these reasons, birth rate dips below death rate and there is a natural decrease in the population.

## Major factors affecting population growth

### Agricultural change

Changes in agriculture occur early in a country's development. Even at intermediate levels, technology improves yields and saves labour. This frees some workers for industry and more rapid economic growth. In the Industrial Revolution in the UK, factories needed a large workforce, so for a while larger families were a benefit. Soon, however, technological advantages reduced the need for labour, making smaller families more desirable and reducing population growth.

### Urbanisation

**Rural–urban migration** is common in poorer countries as cities are believed to have greater opportunities, and generally do. One major reason for such migration is to seek better educational opportunities for children. Children's labour is therefore of less value in cities than in rural areas. Highly urbanised societies tend to have lower rates of population growth.

**Did you know** ??????

A model is a diagram used by geographers to help explain a system or process. It simplifies what is happening in the real world, leaving out detail so that the basic ideas are clear. The demographic transition model is a geographical model, simplifying changes in birth and death rates over time so that the pattern becomes clearer and the factors affecting these changes can be explored.

**Key term**

**Rural–urban migration:** moving home from a rural area to settle in a town or city.

## Education

As levels of educational achievement increase, bringing improved standards of living, children become an economic disadvantage. Fewer children means parents have more money to be spent on each one, giving them better future chances. Many parents in poorer countries see education as their children's best chance in life. Increased educational opportunities tend to result in lower rates of population growth.

## The changing status of women

As economies develop and education improves, opportunities for girls increase alongside those for boys. With development a larger workforce is required, so women must participate more in paid work outside the home. Reaching a good standard of living in a household requires two incomes.

Over time, prejudice against women holding more senior positions at work reduces. Equality increases and is perceived as not only acceptable but desirable. Countries like Sweden have changed the law to increase the proportions of women in management and government. Improvements in the status of women, leading to lower birth rates, slows down the rate of population growth.

However, achieving highly in any career demands a large time commitment, leaving less opportunity for taking maternity leave or caring for children. Some women make deliberate choices regarding not having children or having them later, and these increase as an economy develops. Larger families, or even having a family at all, may be rejected. One in five women in the UK today is childless, compared with one in ten in their mothers' generation. Another issue for working women is childcare, which can be expensive.

## Medical factors

The development of modern medicines has meant that more and more people are kept alive and life expectancy has increased. Similarly the introduction of vaccination and immunisation programmes such as smallpox vaccination has helped people to live longer. Greater numbers of doctors, nurses and hospitals have resulted in better health care in many countries. Cleaner drinking water and better sewage disposal means that a lot more people have access to clean drinking water than before, with a much lower risk of disease and death.

## Political factors

Government policy can be aimed at increasing or decreasing fertility rates. The governments of Italy, Germany and Japan all offered inducements and concessions to those with large families in the 1930s to try to increase the size of the population. Many LEDC governments such as India and China have intervened to reduce fertility rates to control population growth. More recently Malaysia introduced a similar policy. Some countries with ageing populations may try to increase fertility rates with tax incentives to families or actions such as child benefit and maternity and paternity leave.

## Activities

1 Study diagram **A**.

a Describe how birth and death rates change through the five stages of the demographic transition model.

b How do these changes affect population growth?

c Why is Stage 1 really a thing of the past?

d What is likely to happen to the number of countries at Stage 5 in the future?

2 Study map **C**.

a Research the internet for birth and death rates by country and suggest at least two countries for each of Stages 2, 3, 4 and 5.

b Describe the distribution of countries at each stage of the demographic transition model.

3 Imagine you are a factory worker in an urban region of Brazil. Are you more likely to have a small or a large family? Explain your reasons, not forgetting that individuals may hold varying opinions.

4 Why might women in Germany and other Stage 5 countries have a very small family?

## 5.3 How do we use population pyramids?

### Population pyramids

A population pyramid is a type of bar graph used to show the **age** and **gender structure** of a country, city or other area. The horizontal axis is divided into either numbers or percentages of the population. The central vertical axis shows age categories: every 10 years, every 5 years or every single year. The lower part of the pyramid is known as the base and shows the younger section of the population. The upper part, or apex, shows the elderly.

Interpreting population pyramids tells us a great deal about a population, such as birth rates, to a lesser extent death rates, life expectancy and the level of economic development (or stage in the demographic transition model).

### Population pyramids and the demographic transition model

#### Stage 1

The Stage 1 pyramid has a very wide base due to its extremely high birth rate (up to 50 per 1,000 per year). However, **infant and child mortality rates** are high (only 50 per cent may reach their fifth birthday), so the sides of the pyramid curve in very quickly. Death rate is high in all age groups, so life expectancy is low. The result is a very narrow apex and the shortest of all the pyramids.

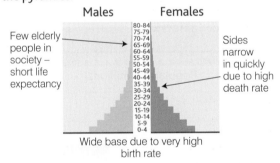

Few elderly people in society – short life expectancy

Sides narrow in quickly due to high death rate

Wide base due to very high birth rate

#### Stage 2

Stage 1 and 2 pyramids are similar in shape. Death rate begins to fall, making the sides of the Stage 2 pyramid slightly less concave. The apex shows a few extra elderly people as life expectancy begins to rise.

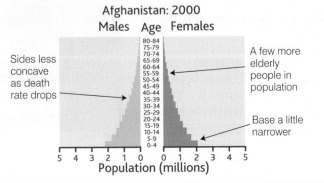

Afghanistan: 2000

Sides less concave as death rate drops

A few more elderly people in population

Base a little narrower

Population (millions)

**A** Population pyramids and the demographic transition model (continued on page 104)

**B** A Stage 5 family

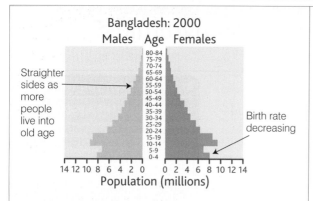

Bangladesh: 2000
Males Age Females

Straighter sides as more people live into old age

Birth rate decreasing

Population (millions)

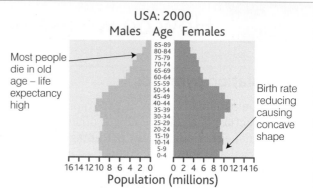

USA: 2000
Males Age Females

Most people die in old age – life expectancy high

Birth rate reducing causing concave shape

Population (millions)

### Stage 3

The narrowing base shows the decrease in birth rate typical of Stage 3 countries. The sides become straighter. Birth rate decreases quickly. Health improvements allow even more people to live into old age.

### Stage 4

This pyramid has become straight-sided, showing a steady low birth rate. High life expectancy allows most people to live into their 60s and 70s and a significant minority into their 80s.

### Stage 5

Germany has been at Stage 5 since the 1970s – one of the first countries to reduce its birth rate to this extent (photo **B**). When today's middle-aged people become elderly, there will be few adults of working age to support them. A Stage 5 population is not sustainable.

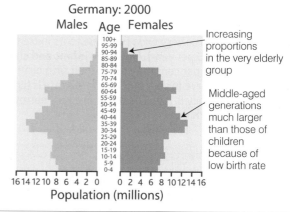

Germany: 2000
Males Age Females

Increasing proportions in the very elderly group

Middle-aged generations much larger than those of children because of low birth rate

Population (millions)

 *Population pyramids and the demographic transition model*

# Future population change

Diagram **D** shows population pyramids for India in 2000, 2025 and 2050. Computer programs can predict future population patterns from today's statistics. Although India's birth rate has fallen, its base remains wide because there are so many young adults in their child-bearing years. By 2025 the number of babies born each year will have stabilised, reducing slightly by 2050. Increasing numbers live into old age, and the 90–94, 95–99 and 100+ age groups are included on future graphs. By 2050 most people will live into their 70s and India will have all the characteristics of a Stage 4 population.

Urban areas of many countries at lesser stages of development are predominantly male in all age groups up to 60–64. Usually more boys are born than girls, explaining the differences at the base (diagram **E**). Rural–urban migration in search of work remains common in countries at lesser stages of development. Men and older boys leave the women, younger children and elderly behind in rural areas. Cities offer greater opportunities to earn money, which can be sent back to improve the family's standard of living (photo **C**). Sometimes whole families make the migration.

**C** *Men queuing for work in Kolkata, India*

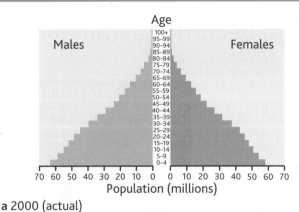

**a** 2000 (actual)

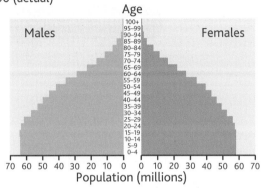

**b** 2025 (predicted)

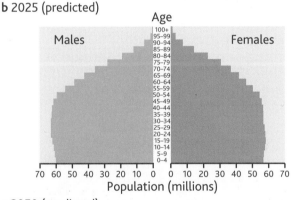

**c** 2050 (predicted)

**D**  *Population pyramids for India*

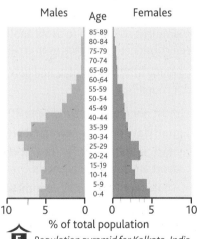

**E**  *Population pyramid for Kolkata, India*

**Activities**

1  Study diagram **D**.

a  Describe the shapes of the three pyramids for India, making the similarities and differences clear.

b  Explain in your own words how India's population pyramids are likely to reflect economic changes over this period of time.

2  Study diagram **E**. Label a copy of Kolkata's population pyramid to show its unusual characteristics.

3  Research the pyramids for two countries at different stages of development in the early 21st century. Try to predict likely changes in population and development in the future.

**∞links**

India's population pyramids can be used to predict future changes in the country, both demographically and economically. The same can be done for other countries.

See **http://populationpyramid.net**.

# 5.4 How can a population become sustainable?

A **sustainable population** is one whose growth and development is at a rate that does not threaten the success of future generations. Countries at Stage 4 of the demographic transition model, with low birth and death rates, are the most sustainable. The economy is stable or growing and the standard of living is maintained or improving.

Stage 5 populations are not sustainable because numbers are decreasing. In Japan, calculations were done to predict how long it would take the country to die out if current low birth rates continued well into the future. Although it is highly unlikely to happen, the fact that it was even thought about was worrying.

> **In this section you will learn**
>
> the reasons for the one-child policy in China from 1979
>
> the severity of the rules imposed
>
> how the one-child policy has changed over recent years and the outcomes of these changes.

## Case study

## Will China have a sustainable population?

### The early days of the one-child policy

During the 1970s the Chinese government realised that the country was heading for famine unless severe changes were made quickly. Change to an industrial economy at the expense of farming had already caused a catastrophic famine from 1959 to 1961, with 35 million deaths. A 'baby boom' followed and population was growing too fast to be sustainable. The government stepped in to avoid another crisis. However, its methods have been considered too strict, even cruel.

Beginning in 1979, the one-child policy said that each couple:

- must not marry until their late 20s
- must have only one successful pregnancy
- must be sterilised after the first child or must abort any future pregnancies
- would receive a 5 to 10 per cent salary rise for limiting their family to one child
- would have priority housing, pension and family benefits, including free education for the single child.

Any couples disobeying the rules and having a second child were severely penalised:

- a 10 per cent salary cut was enforced
- the fine imposed was so large it would bankrupt many households
- the family would have to pay for the education of both children and for health care for all the family
- second children born abroad are not penalised, but they are not allowed to become Chinese citizens.

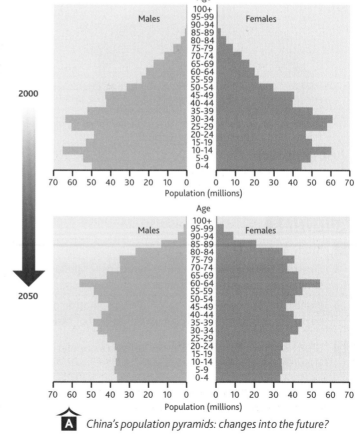

**A** China's population pyramids: changes into the future?

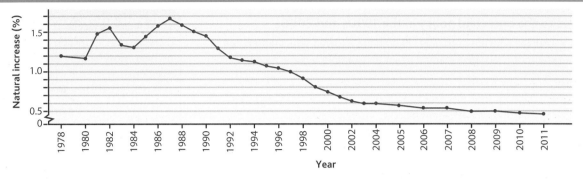

**B**   *China's population growth through the one-child policy*

Pressure to abort a second pregnancy even included pay cuts for the couple's fellow-workers so they would make life unbearable. The 'Granny Police' – older women of the community entrusted with the task of keeping everyone in line – kept a regular check on couples of childbearing age, even accompanying women on contraception appointments to make sure they attended.

China is racially mixed but over 80 per cent of the population is of the Han race. Minority groups could become unsustainable under the one-child policy, so they were exempt. In rural areas, where sons are essential to work the family land, a second pregnancy was allowed if the first child was a girl, in the hope of getting a boy.

### The problems and benefits of the policy

The one-child policy has been controversial for many reasons:

- Some women were forced to have abortions as late as the ninth month of pregnancy.
- Women were placed under tremendous pressure from their families, workmates, the 'Granny Police' and their own consciences and feelings.
- Local officials and central government had power over people's private lives.
- Chinese society prefers sons over daughters. Some girls were placed in orphanages (photo **C**) or allowed to die (female infanticide) in the hope of having a son the second time round.
- Chinese children have a reputation for being over-indulged because they are only (single) children, hence the name 'Little Emperors' (photo **D**).

China's one-child policy has brought important benefits to the country, however. The famine previously forecast has not happened. Population growth has slowed down sufficiently for people to have enough food and jobs. It is estimated that 400 million fewer people have been born. Increased technology and exploitation of resources have increased standard of living for many. New industries have lifted millions out of poverty. This is partly the result of the one-child policy, but new technology from other countries has helped.

**C**   *Children in a Chinese orphanage*

**D**   *Only children are often spoiled*

## Changes to the one-child policy in the 1990s and 2000s

- Young couples who are both only children are allowed two children, but government officials are expected to set an example by only having a single child.

- As people have become wealthier, some are choosing to have larger families and simply pay the fines.

- Couples no longer need to obtain permission to have a first child.

- The Shanghai government is considering introducing incentives to encourage couples to have two children.

- It is now illegal to discriminate against women who give birth to baby girls and has prohibited sex-selective abortions.

- In 2008, following a devastating earthquake that killed lots of children, the policy was relaxed in Sichuan Province.

## Consequences of the one-child policy

- With less time needed for childcare, women have been able to concentrate on having a career.

- The policy will not change until at least 2015. In 2008, China still had 1 million more births than deaths every five weeks and half the population lived on less than $2 a day.

- With boys being favoured rather than girls, China has a serious gender imbalance. There are now 60 million more young men than young women.

### Did you know ??????

There are over 15 million orphans in China, most healthy girls, abandoned as a result of the one-child policy and Chinese society's economic and social preference for boys. Missionary-run orphanages are usually very good, but in state institutions the girls are neglected and sometimes treated cruelly.

### Did you know ??????

Chinese couples had to apply for permission to try to become pregnant and take their turn. If they did not succeed in a six-month period, they had to wait for another opportunity.

---

We cannot just be content with the current success. We must make population control a permanent policy.

*Adapted from the* People's Daily *(China's Communist Party newspaper), 2000*

Beijing mother-of-one, Zhao Hui, who has a four-year-old daughter called Zhang Jin'ao, says she never wanted more than one child. 'One child is enough. I'm too busy at work to have any more,' says the 38-year-old.

*Adapted from BBC News website* **news.bbc.co.uk***, 20 September 2007*

**E** *News items about the policy*

## Activities

**1** Study diagram **A**.

a  Describe the shape of each pyramid.

b  On copies of the pyramids, label the key features, paying particular attention to any similarities and differences.

**2** Study graph **B**.

a  Describe the shape of the graph.

b  Do you think the one-child policy has been a success in controlling population growth? Explain your answer.

**3** Study the text above.

a  Make a list of the controversial aspects of the one-child policy. Remember that the Chinese government and people may not see everything in the same way as outsiders.

b  What were the successful outcomes of the policy during the 1980s?

**4** Make a decision – was the Chinese government wise to introduce the one-child policy? Have the benefits outweighted the disadvantages? There is no right or wrong answer here – you simply need to justify your opinion clearly.

## 5.5 How can population growth be managed?

Case study

## Transmigration in Indonesia

Indonesia is a rapidly developing country in south-east Asia (map **A**). It comprises a group of some 17,500 islands (known as an archipelago), two of the largest being Sumatra and Java. With a population of about 240 million, it is the 4th most populous country in the world. The population is growing at a rate of just over 1 per cent per year, with the birth rate at 18.1 per 1000 (2011) and the death rate at a very low 6.3 per 1000 (2011). Look back to diagram A (page 99) to consider what stage of the demographic transition model Indonesia is in.

**A** Location of Indonesia

### In this section you will learn

how Indonesia has tried to manage population growth by moving people from overpopulated to underpopulated areas

other ways of coping with population pressures.

### Key terms

**Transmigration**: a population policy that aims to move people from densely populated areas to sparsely populated areas and provide them with opportunities to improve the quality of their lives.

**Industrialisation**: a process, usually associated with the development of an economy, where an increasing proportion of people work in industry.

## What is transmigration?

Whilst population growth is now slowing down in Indonesia, back in the 1950s and 1960s, it was growing rapidly and the islands of Java, Bali and Madura were in danger of becoming over-populated. This occurs when there are not enough resources (for example, food, water, jobs and housing) to adequately support the number of people.

In 1969, in an attempt to re-distribute the population, the Indonesian government embarked on an ambitious project called **transmigration**. This involved encouraging people to move from densely populated islands such as Java to the more sparsely populated outer islands, such as West Papua (Irian Jaya). Transmigration was a chance for people to escape from the poverty of overcrowded urban slums to become land owners elsewhere and earn money through farming. Between 1979 and 1984, at the height of the programme, more than 2.5 million people were involved. In total, over a million people have resettled in West Papua (Irian Jaya) alone.

Financed by the World Bank and the Asian Development Bank, transmigration continued through the 1980s but recent financial difficulties and changes in government have led to the policy being scaled down.

**B** Transmigration settlement near Geumpang, Aceh in Indonesia

## Recent developments

In 2006, an estimated 20,000 families took advantage of the transmigration programme, supported by the Department of Manpower and Transmigration. By 2010, there were about 250,000 families who wanted to move, the government was only able to sponsor 10,000 families at a cost of some $US160.5 million.

Following the eruption of Mount Merapi in 2010, the government offered to re-locate tens of thousands of displaced people from Java to Kalimantan. By 2011, some 2,000 families had taken up the offer encouraged by free transportation and the promise of two hectares of land and living costs for six months.

## What have been the effects of transmigration?

Whilst transmigration has probably eased overcrowding in some towns and cities, the policy has led to problems.

### Economic

■ Rather than reducing poverty, critics suggest that transmigration has simply re-distributed poverty.

■ Many new migrants lacked the necessary farming skills to make productive use of their new land and some abandoned their new homes to become refugees.

■ Settlements were often poorly planned, with few shops, roads and services such as water, sanitation and electricity (photo **B**).

■ Re-settlement was extremely expensive, costing $US7,000 per family in the 1980s.

### Environmental

■ Transmigration has been blamed for accelerating the rate of deforestation in previously sparsely populated regions.

■ Poor land use practises, such as over-cultivation, have led to issues of soil erosion.

### Social

■ Clashes have occurred between migrants and the indigenous populations, particularly in more remote islands. In 2001, hundreds died when the local Dayaks and the transmigrant Madurese (from Madura) clashed.

■ Traditional land rights were often ignored as land ownership was granted to the new settlers.

■ There have also been religious clashes between the Islamist migrants and the largely Christian local people.

### Political

■ Some critics have suggested that transmigration was encouraged by the government primarily to increase national security and control indigenous people in the outer islands.

Despite the problems, the transmigration policy has resulted in the re-settlement of an estimated 20 million people, mostly to Sumatra and Kalimantan. Whilst the population pressures have continued to increase in cities such as Jakarta, the situation would probably have been much worse if transmigration had not occurred.

## Industrialisation: another approach to coping with population pressures

For the future, as birth control programmes have led to a slowing down of population growth in Indonesia (families are encouraged to have two children only), the transmigration programme will probably continue to decline. In its place, as Indonesia expands its programmes of resource exploitation (for example, minerals, timber, oil palm and shrimp farming) and **industrialisation**, more and more poor people will migrate to the remoter islands looking for work. Since the 1990s the government has encouraged industrialisation in the remoter islands through its 'Eastward Development Policy'. By encouraging resource development and industrialisation, workers will naturally be attracted. This can be considered another approach to coping with a rapidly growing population.

### Activities

1   Study the information in the case study and answer the following questions.

a   What is transmigration?

b   How does transmigration address the issue of rapid population growth?

c   What are the advantages of the transmigration programme to the migrants?

d   How have local indigenous people been affected by the influx of new migrants?

e   Do you think transmigration has been a success? Explain your answer.

2   Use the information in this case study together with your own internet research to find out more about the environmental impacts of the transmigration programme. You might choose to focus on the Mega Rice Project in Kalimantan. Use maps and photos to support your study. Assess the success of the transmigration programme – did the economic benefits outweigh the environmental damage?

# 5.6   What are the issues and opportunities for an ageing population?

Populations in richer countries are ageing. Low birth rates and smaller families result in fewer children and adolescents. Better health care and more advanced medicines allow people to live longer, increasing the proportion of elderly people. Today's older people in countries at further stages of development tend to be wealthier, fitter and have wider interests, with the spare time to enjoy these. This is, however, the age group with the most expensive needs, especially in terms of health care. The 85+ age group – the very elderly – is growing fastest, increasing stress on health and social welfare systems (photo **A**).

**A**   *The 85+ age group is growing fast. This couple is living in a care home*

**In this section you will learn**

the ways in which an ageing population is different from younger populations

the demands on a country that has an ageing population

how, as the proportion of elderly people increases, the costs to the government increase dramatically

different ways of solving the problem of an ageing population

how to evaluate different approaches to coping with an ageing population.

## The issues

### Health care

The demand for health care increases because more illness occurs in old age. The elderly visit their GP (doctor) more often. They have more hospital appointments and more time in hospital than younger or middle-aged people. The government has to find more funding to support older people and this comes from taxation of present workers.

### Social services

Elderly people need other services such as nursing homes, day-care centres and people to help them to care for themselves at home. These special needs put financial pressure on a country.

### The pensions crisis

Life expectancy is higher in developed countries than in developing countries. In wealthier countries, people expect to be able to retire from work and have a pension (an income) for the rest of their lives. As there are more elderly people and the proportion of working people is decreasing, so the taxes must increase to pay the pensions bill.

The state pension began in 1908 when male life expectancy was 67 and retirement age was 65. The average person would therefore receive their pension for only two years. Today, the situation could not be more different as life expectancy has risen to 80. In response to this, the retirement age is set to increase over the next few years.

**Did you know** ? ? ? ? ? ?

With the second lowest birth rate in Europe, Italy faces population decline in the 21st century. Projections suggest that 40 per cent of Italians will be over 60 and only 15 per cent under 20 by the end of this century.

## Activity

**1**  Study diagram **A** on pages 103–104.

  a  Describe how the numbers and proportions of elderly people change as a country passes through the demographic transition model.

  b  Look at the pyramids for Stages 4 and 5. What has happened to the gender ratio?

## The opportunities

The situation has a positive side too. Younger retired people (those in their later 60s and early 70s) contribute a great deal to the economy (photo **B**). They are relatively wealthy and have lots of leisure time. They spend money on travel and recreation, providing jobs in the service sector. Many do voluntary work and some still do paid work and therefore pay taxes.

**B**  Younger retired people contribute a lot to the UK's economy

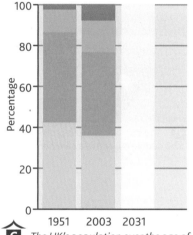

Key
- ■ 85 and over
- ■ 75 – 84
- ■ 60 – 74
- ■ 50 – 59

**C**  The UK's population over the age of 50, 1951–2003

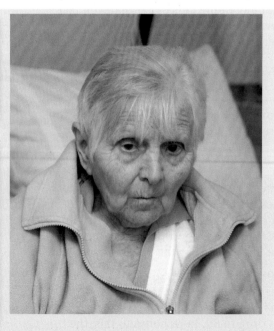

As soon as my hair turned white, people started to ignore me. It was as if I had become invisible.

*Anna, 90, and still coping in her own home with no outside help*

Britain's elderly are being neglected, poorly treated and marginalised by the country's health system. Inspectors found that many older patients were hungry because meals were taken away before they could eat them.

*The Independent, 27 March 2006*

It's horrid having to rely on other people for the most basic things. It's bad enough when it's your own family and friends, but when it's strangers from social services it's so much worse.

*Barry, 95, who has four carers every day to get him up, washed, dressed and, later, back to bed*

How bad is the UK's pension crisis? Every day seems to bring fresh warnings that Britons will not have enough money to live on when they retire.

*Adapted from BBC News website news.bbc.co.uk, 23 September 2005*

**D**  *Quotes on being elderly, pensions and the health crisis*

## ■ Ageing in the EU

Table **E** shows **European Union (EU)** birth rates are very low – in fact, they have never been so low and they may decrease even more. Smaller families and later motherhood could soon result in a noticeable decline in population in some countries. In each generation there are fewer parents, so fewer children are born. In 2008, 1.6 babies were born in the EU for every woman, but 2.1 are needed for a population to be sustainable. Western European birth rates are higher than those in Eastern Europe, but Germany – a Western European country – has an even lower birth rate (table **E**).

**E**  *Selected EU birth rates (2011)*

| Country | Birth rate (per 1,000) |
|---|---|
| Ireland (W) | 16.1 |
| France (W) | 12.3 |
| UK (W) | 12.3 |
| Netherlands (W) | 10.2 |
| Poland (E) | 10.0 |
| Bulgaria (E) | 9.3 |
| Latvia (E) | 10.0 |
| Germany (W) | 8.3 |

The average global birth rate was 19.95 per 1000 in 2009.

Note:  (W) Western European country
       (E) Eastern European country

### Case study

### France's solution

France is tackling its ageing population with a strong **pro-natal policy** – it encourages people to have children to produce a more favourable age structure and **dependency ratio**. This has had some effect, but has not been entirely successful as graph **F** shows.

Couples are given a range of incentives to have children:

■ three years of paid parental leave, which can be used by mothers or fathers

■ full-time schooling starts at the age of three, fully paid for by the government

■ day care for children younger than three is subsidised by the government

■ the more children a woman has, the earlier she will be allowed to retire on a full pension.

Nicole Falcou is 53. She lives in Muret, close to Toulouse in south-west France. She is married with three daughters aged between 14 and 22 and works locally with disabled children. She has three children of her own, so she is entitled to extra benefits from the government, including retiring in her early 50s on a full state pension if she chooses. Although a pension is always less than a salary, as her children grow up and become less financially dependent this is a tempting offer.

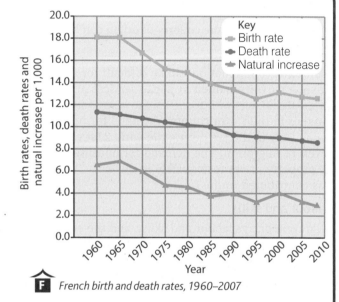

**F**  *French birth and death rates, 1960–2007*

### Activities

2   Study graph **C** on page 112.

a   Copy and complete the graph, adding a similar bar for 2031 by using the data in table **G**.

b   Describe the changes shown by the bars between 1951 and 2031.

c   Explain why these changes are happening.

3   Study graph **F**.

a   Describe the pattern of the birth rate line.

b   List the problems affecting a country with a very low birth rate.

**G**  *UK population aged over 50, 2031*

| Age group | % |
|---|---|
| 50–59 | 28 |
| 60–74 | 44 |
| 75–84 | 20 |
| 85 and over | 8 |

### Key terms

**Dependency ratio:** the balance between people who are independent (work and pay tax) and those who depend on them. Ideally, the fewer dependents for each independent person, the better off economically a country is. Here is the formula (figures can be in numbers or percentages):

$$\frac{\text{number of dependent people}}{\text{number of independent people}} \times 100$$

# 5.7 What are the impacts of international migration?

## Push–pull factors

People move home for many different reasons. Every individual's decision to move is the result of **push–pull factors**. Negative aspects of a person's home area push them away from it and make them look for somewhere better. Positive characteristics of new places, which attract people to move there, are called pull factors.

## Impacts of international migration

The impacts of **migration** on the **country of origin (source country)** and the **host country (receiving country)** can be positive or negative.

### Economic

Migrant workers often send money back to their country of origin to help their families. This means that money leaves the host economy – a disadvantage – but the country of origin can benefit enormously.

### Housing

Finding accommodation can be difficult for migrants in the UK. Some have been helped by social services and this causes resentment from UK citizens who feel they are being treated as second-class citizens in their own country. Demand for housing has grown immensely in the UK during the early 21st century, fuelled by high levels of migration. The demand for housing is greater than supply, so property prices rose quickly in the early 2000s and immigration has contributed to this.

### Labour and skills

Migration brings labour and skills, and the economies of the UK and the EU have grown as a result. Most migrants are more successful than they would have been at home, although some are less fortunate. Exploitation does happen and not everyone earns as much as they had expected. Tragedies have occurred when gang masters, who often control large numbers of workers in agriculture and shellfish harvesting, have been negligent. The deaths of 23 Chinese cockle-pickers in Morecambe Bay in February 2004 was perhaps the worst example.

### Social

Too many migrants can be a burden. Schools taking many **immigrant** children may be under pressure. British parents sometimes feel this reduces opportunities for their own children because teachers are too busy with those whose first language is not English. On the other hand, cultural mixing is often seen as positive as long as racial prejudice does not become a problem.

---

### In this section you will learn

the concept of migration and people's reasons for moving home (push–pull factors)

the positive and negative impacts of international migration

who moves within the EU and why

who comes to the EU and their reasons for wanting to live here

the differences between voluntary economic migrants and refugees

the benefits and difficulties of international migration for EU countries.

### Key terms

**Push–pull factors**: push factors are the negative aspects of a place that encourage people to move away. Pull factors are the attractions and opportunities of a place that encourage people to move there.

**Migration**: the movement of people from one permanent home to another, with the intention of staying at least a year. This move may be within a country (national migration) or between countries (international migration).

**Host country (receiving country)**: the country where a migrant settles.

**Country of origin (source country)**: the country from which a migration starts.

**Immigrant**: someone entering a new country with the intention of living there.

## A Slovak girl working in Sussex

Jana Susinkova came to the UK in 2002 with her Czech boyfriend. She was only 18 and he a little older. She worked as a domestic cleaner, undercutting the level local women charged by at least £1 per hour. She had enough work to keep busy six days a week. Her boyfriend was a mechanic and odd-job man, and his job provided accommodation for them.

Late in 2007 Jana returned home to Slovakia. Her boyfriend had already left to find a job in the Czech Republic, where the growing economy offered increasing opportunities for skilled people. While in the UK they had saved enough money to buy the materials and labour to build a four-bedroom house in the Czech Republic, giving them an excellent start to their married life. Jana's English had become fluent, so she quickly found a well-paid job where she uses it every day.

**A**  *Working as a cleaner*

**B**  *Push–pull factors for a family in West Africa*

| Family member | Situation |
|---|---|
| Father | Subsistence farmer – crops unpredictable due to climate. Part-time fisherman – catches are reducing because of overfishing. |
| Mother | Housewife with limited primary education. |
| Adult son | Secondary education completed. Would like the chance to go to university or obtain an interesting job. |
| Daughters | Part-way through school. Want to get as well qualified as possible. School resources are sometimes in short supply. |

## Activities

**1** Write the factors listed below into two groups of push and pull factors in terms of someone's possible migration:

job prospects    natural disasters

low income    housing shortages

health care    intolerance

high standard of living    high unemployment

improved housing    racial/religious tolerance

attractive environments    high wages

political or social unrest

educational opportunities    difficult climate

**2** Study table **B**, which shows the factors affecting a family in West Africa who are considering moving to the EU. The family consists of two parents, an adult son and two younger daughters who are still at school.

a The father is keen to move but the mother wants to stay. Suggest reasons for these opinions.

b Identify the push and pull factors affecting the children.

c If the family moves to the EU how might the individual members benefit?

d Suggest any possible problems or issues that the family might face.

## Migration from outside the EU

Europe currently receives over 2 million immigrants from beyond its borders a year – more than any other world region. The ratio between current population and immigrants is higher for Europe than for the USA. European population is changing more in age and racial structure due to immigration than by changes in birth and death rates.

About 9.4 per cent of the EU's people are foreign-born, compared with 10.3 per cent in the USA and almost 25 per cent in Australia. The range of countries around the world from which migrants to Europe come has changed. Africa and Asia are the major sources of immigrants, but not the same regions as previously.

## Migration within the EU

There are two categories of migrants within the EU: those moving between countries and those coming in from beyond the borders. Wealthier countries usually receive immigrants searching for work and a better lifestyle. Poland and other Eastern European countries joined the EU in 2004. Since that date, many people have moved temporarily or permanently to the UK and other Western EU countries for work.

---

**Case study**

## Polish workers migrating to UK

Since 2004, 1.5 million Eastern Europeans have entered the UK, two thirds of whom are Polish. Most have found formal jobs with much better pay than they would receive at home. Polish workers in the UK earn on average five times as much as they would at home, while the UK cost of living is only twice that in Poland. Most migrants pay tax, which contributes to the UK's economy. However, some work in the informal economy – working for cash and not paying tax. Polish workers also use UK health and education services, which add to the government's costs.

Overall, the UK economy has benefited from the influx of migrants from Poland. Poles are now the most numerous of foreign nationals in the UK. There are 545,000 Polish passport holders living in the UK compared with 75,000 in 2003, the year before Poland joined the EU.

| **D** Occupations of Polish workers in the UK, 2008 ||
|---|---|
| **Occupation** | **Total** |
| Administration, business, management | 33% |
| Hospitality and catering | 22% |
| Agriculture | 10% |
| Manufacturing | 8% |
| Health service | 6% |
| Food processing | 5% |
| Retail | 5% |
| Construction | 5% |
| Others | 6% |

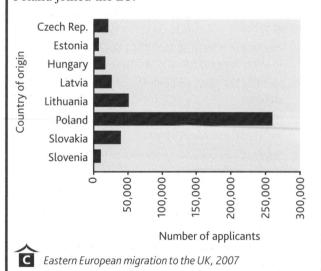

C  Eastern European migration to the UK, 2007

**Activities**

3  Study graph **C** and table **D**.

a  Using figures, describe the dominance of Poland in EU migration to the UK.

b  Use the information in table **D** to draw a pie chart to show the occupations of Polish workers in the UK.

c  Discuss the types of occupations Polish workers undertake. Are there any trends?

4  Suggest the benefits and difficulties of migration within the EU. Who benefits and who does not?

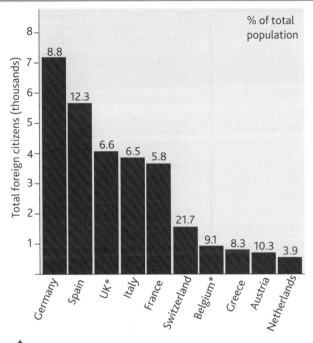

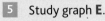
**E**  'Top ten' EU countries with the highest total of foreign citizens (2009)(* UK and Belgium 2008)

**F**  Passport control at Heathrow Airport – a popular point of entry for economic migrants

## Labour migration

Cheaper travel and more information attract skilled and unskilled labour to Europe. Many EU residents would like the flow of migration to reduce, but the United Nations predicts that immigration into the EU will rise by 40 per cent over 40 years. Immigration is a subject of political debate in all EU countries. Spain's immigrant population grew by 400 per cent in 10 years in the early 21st century. Italy expects 100,000 Romanians in the years following Romania's joining of the EU. Although Italy needs workers, not everyone is happy with such a large influx of new people.

Europe needs immigrants because of its falling birth rate and the resulting lack of workers. Highly skilled workers often come to the EU to take temporary jobs in areas of shortage such as teaching, nursing and high-tech computer jobs. About 20 per cent are graduates. Nevertheless, many people see immigrants as a problem rather than as an opportunity.

**Study tip**

Proportional symbols are drawn to scale to show the values they represent visually. They need careful construction. One of the most common types is proportional circles. The best method is to make the radius of the circle proportional to the value.

## Activities

**5**  Study graph **E**.

**a**  On a sheet of plain paper, draw proportional circles to scale to show the total foreign citizens as a percentage of total population. Cut them out carefully.

**b**  On a blank outline map of Europe, stick each symbol on or beside the country it represents.

**c**  Give your map a title and explain the scale of your circles in a key.

**d**  Describe the pattern shown in your completed map.

Case study

# International labour migration to the EU: Senegal to Italy

Senegalese children love football and support home teams, but many also support Lazio, AS Parma and other Italian teams. The reason for this is that many of these children's fathers and brothers already work in Italy. Patterns of emigration from Senegal are well established and many young men are keen to work in Italy.

**G**  *Many Senegalese fishing villages have been abandoned*

Opportunities are limited in Senegal, where many people are subsistence farmers. With limited services and high rates of urban unemployment, young men are keen to emigrate to the EU to find work. Money is often sent home to support children's education or improve living conditions. For example, in the villages of Beud Forage money has been used to set up water and electricity supplies.

## Refugee movements to the EU

**Asylum seekers** are people who are at risk if they stay in their own country. They become refugees when they settle in another country. One-third of EU immigrants claim to be asylum seekers. Since EU countries have reduced the number of EU migrants they would allow in, some **economic migrants** have claimed to be asylum seekers, believing this would give them a better chance of being allowed to stay in the EU. Unfortunately, this has sometimes caused strong feelings against genuine asylum seekers.

Despite being criticised by the United Nations for not taking enough genuine refugees, the EU has a good record of accepting those displaced by war. The 1990s Bosnian war produced hundreds of thousands of refugees to the EU. Germany alone took 400,000, many of whom returned home once the situation was peaceful.

Today, the wars in Iraq and Afghanistan provide most asylum claims. Two million Iraqis have already left the country, some for neighbouring countries and some to the EU. Another 1.8 million refugees live away from their homes in Iraq and many feel sufficiently threatened to want to leave.

Sweden is particularly generous to asylum seekers (table **H**). By 2007, 70,000 Iraqis already lived there – half of those coming to the EU. The Netherlands, Germany, Greece, Belgium and the UK have given homes to most of the rest. With the Iraq war coming to an end, asylum requests to EU countries from Iraqis have decreased to 19,176 in 2010 from a peak of 38,286 in 2007. Afghans are now the largest group seeking refuge in the EU with 22,939 asylum requests in 2010.

**H**  *Annual asylum applications (2010)*

| EU country | Number of applications |
|---|---|
| Sweden | 31,820 |
| Netherlands | 13,330 |
| Belgium | 19,940 |
| Germany | 41,330 |
| Denmark | 4,970 |
| Ireland | 1,940 |
| UK | 22,090 |
| Spain | 2,740 |

## Activities

6  **a**  Who are 'asylum seekers' and how do they differ from economic migrants?

**b**  Why do wars often produce large numbers of asylum seekers?

**c**  Represent the data in table **H** in the form of bars on an outline map of Europe.

**d**  Why do you think the countries in table **H** are popular destinations for asylum seekers?

**e**  Use the internet to make your own study into asylum seekers associated with a recent world conflict, such as civil war in Libya (2011).

## Key terms

**Asylum seekers**: people who believe that their lives are at risk if they remain in their home country and who seek to settle in another (safe) country.

**Economic migrant**: someone trying to improve their standard of living, who moves voluntarily.

## ∞ links

To investigate further why Eastern European migrants are coming to the UK, visit **www.migrationwatchuk.com**.

# 6 Contemporary issues in urban settlements

## 6.1 What are the characteristics and causes of urbanisation?

In 2008, for the first time in history, over half of the world's population lived in towns and cities. **Urbanisation** – where there is an increasing proportion of city dwellers in contrast to those in the countryside – is a worldwide process. It began at different times in different parts of the world and occurred at contrasting paces, as it continues to do so today (graph **A**). The contrasting proportions of people living in towns and cities is shown in map **B**.

### In this section you will learn

the process of urbanisation and how it varies throughout the world and over time

the causes of urbanisation.

### Key term

**Urbanisation**: a process where an increasing proportion of the population lives in towns and cities resulting in their growth.

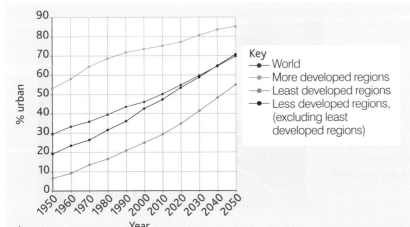

**A** Urban population, 1950–2050

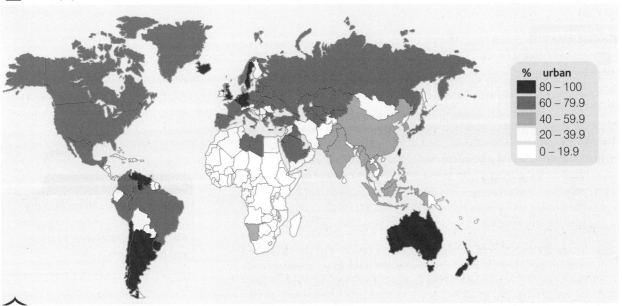

**B** Global urban population, 2000

## Causes of urbanisation

There are two causes of urbanisation: rural–urban migration and natural increase. The initial cause of urbanisation is usually rural–urban migration. This is the result of push–pull factors, as shown in the photos in **C**. The people that migrate into the towns and cities are generally young and this results in relatively high levels of natural increase. The high proportion of young adults results in high levels of births. Improvements in medical care mean that babies are more likely to survive. This can lead to an increase in the urban population while birth rates remain high.

" Recent droughts have meant yields have been lower and lower, so I've struggled to provide enough food for my family to eat. I have only a small amount of land and can't afford to buy any fertilisers. "

" I came here because I thought we would be better off. I believed we would have a better house and I would have a good job and some money. My children would be able to go to school and we would get better medical help if we are sick. "

**C** *Reasons why people leave the Ethiopian countryside*

## Activities

1. Study graph **A**.
   a. Describe the trends shown by each of the lines. Support your answer with evidence from the graph.
   b. Give examples of how the process of urbanisation has been important at different times and has occurred at different speeds.

2. Study map **B** and a political map in an atlas. Describe the pattern shown by the map. Try to select areas where rates of urbanisation are very high, high, average, low and very low. Give a clear overview of the locations of places that generally fall into such categories. Can you see any exceptions to the patterns you have described?

3. Study the photos in **C**. Imagine you are a resident of the rural area shown in the photo. Relatives have come back to your village and described the city shown in the second photo to you. Explain why you really want to move there.

**Study tip**

Ensure that you can explain the process of rural–urban migration with specific reference to push and pull factors.

# 6.2   How does land use vary in an urban area?

In all towns and cities the **land use** varies. In some areas shops and services may be dominant, whereas other areas may be dominated by housing, industry or recreation. Some areas have mixed land use where the **function** varies. In British cities, certain areas in similar locations tend to have similar characteristics. For example, the central area tends to be the shopping area or **central business district (CBD)**. The area around this is likely to have originally been the oldest part (although more recent building may have changed this) and this is referred to as the **inner city**. Newer areas are generally found on the outskirts, often known as the **outer city or suburbs**. The photos in **A** show the key areas and their features.

> **In this section you will learn**
>
> how urban areas have a variety of functions
>
> the characteristics and locations of some urban areas.

**a CBD**

**b Redevelopment**

**c Suburbs**

**d Inner city**

**A** *Characteristics of different parts of Sheffield*

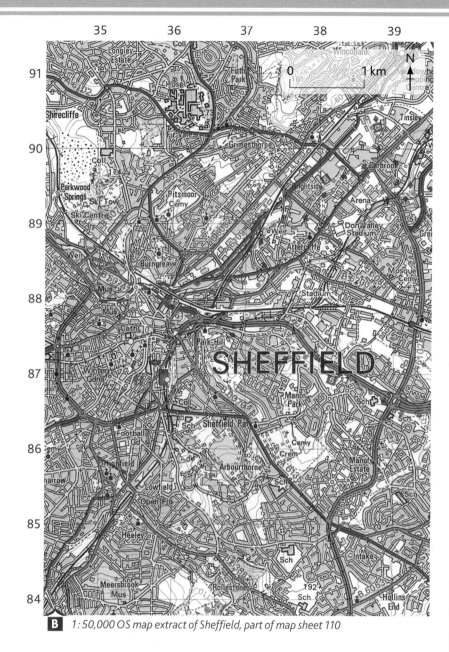

**B**  1 : 50,000 OS map extract of Sheffield, part of map sheet 110

## Activity

1  Study the photos in **A** and map **B**.

a  What is the evidence that photo **a** shows a CBD?

b  Describe the CBD (roughly in the centre of the area within the ring road) from the OS map. You should include reference to its size, shape and functions.

c  Using the photo of the inner city, draw a labelled sketch to show the characteristics of the houses and the environment.

d  Provide evidence to support the idea that land use in Sheffield varies. You should try to give five different points.

# 6.3 What are the issues for people living in urban areas in richer parts of the world?

There are many issues in towns and cities, including those relating to housing, traffic, services and provision for a mixed community. The photos in **A** highlight some of the issues.

 *Issues in urban areas in the UK*

## Activity

1 Study the photos in **A**. Work in pairs.

a For one of them, write three questions to identify key features shown in the photo. Your partner should do the same for the other photo. Your questions should be designed to obtain clear, detailed and thoughtful answers.

b Swap questions. Answer the questions your partner has written on the photo. You should give full and detailed answers to the questions.

c Swap your answers and read and discuss them. Do you agree with the answers given? How good were the questions asked? How could they be improved?

d Together, summarise the issues you think are shown.

## Issue 1: housing

Population in the UK has increased by 10.5 per cent since 1971 and this rate of growth is predicted to continue, giving a population of 65 million by 2025. The number of **households** has risen by 30 per cent since 1971. Most of this increase is because more people live alone – some 7 million (12 per cent in 2009) of the UK's population. New single-person households account for 70 per cent of the increased demand for housing. This is due to people leaving home to rent or buy younger than previously, marrying later, getting divorced and living longer. A third of single-person households are aged over 65.

The government target is to build 240,000 new houses every year by 2016 so that house prices do not spiral out of control as a result of a shortage. Many of these new homes will be built throughout existing towns and cities, with a target of 60 per cent to be built on **brownfield sites** – areas that have been previously built on, usually in the inner city.

### In this section you will learn

the issues facing people living in urban areas in MEDCs, including housing, traffic and economic decline

the success of strategies introduced to manage problems in the CBD and inner city.

### Did you know ??????

Three blocks of high-rise flats in Everton, Liverpool were nicknamed 'The Piggeries' because of the living conditions there.

### Key terms

**Household:** a person living alone, or two or more people living at the same address, sharing a living room.

**Brownfield sites:** land that has been built on before and is to be cleared and reused. These sites are often in the inner city.

**Greenfield sites:** land that has not been built on before, usually in the countryside on the edge of the built-up area.

**City Challenge:** a strategy in which local authorities had to design a scheme and submit a bid for funding, competing against other councils. They also had to become part of a partnership involving the local community and private companies who would fund part of the development.

However, some housing will inevitably be built on **greenfield sites** – areas that have not previously been built on, usually on the edge of the city. Table **B** summarises the points in favour of each of these two alternatives. The photos in **C** show the different types of housing being built to meet the growing demand and contrasting needs of the population in different parts of the city.

**B**   *Advantages of building on brownfield and greenfield sites*

| Advantages of building on brownfield sites | Advantages of building on greenfield sites |
|---|---|
| Easier to get planning permission as councils want to see brownfield sites used | New sites do not need clearing so can be cheaper to prepare |
| Sites in cities are not left derelict and/or empty | No restrictions of existing road network |
| Utilities such as water and electricity are already provided | Pleasant countryside environment may appeal to potential home owners |
| Roads already exist | Some shops and business parks on outskirts provide local facilities |
| Near to facilities in town centres, e.g. shops, entertainment and places of work | Land cheaper on outskirts so plots can be larger |
| Cuts commuting | More space for gardens |

a  Riverside apartments, Bridgewater Place, Leeds

b  Gentrified housing, Cambridge

c  Retirement bungalows, Scartho, Grimsby

d  Family homes, Scartho Top, Grimsby

**C**   *Types of housing*

## Activities

**2**  Read the text and study table **B** on page 124.

a  Why is the demand for housing likely to increase significantly in the next few years?

b  What is meant by the term 'brownfield site'?

c  What are the advantages and disadvantages of building new houses on a brownfield site?

d  Identify a brownfield site that is being used for housing in your local area.

e  What is a 'greenfield site'?

f  Why are many people concerned about the use of greenfield sites for building houses?

g  Is there a greenfield site near you that is being developed for housing? If so, has it caused concern among local people? Explain your answer.

**3**  Study the photos in **C** on page 124.

a  In groups, take on the following roles:

■ professional accountant working in a city-centre office, aged 26 and single

■ a 68-year-old pensioner, widowed and living alone

■ a couple seeking an old, modernised house as a first home together

■ a family with two children aged 10 and 12.

b  For each role, identify which of the four houses shown in **C** would be most suitable and explain the advantages of living there.

## ■ Issue 2: traffic

As we demand greater mobility and accessibility, the number of cars has increased, as has the problem of traffic congestion. More people have more money and welcome the door-to-door service that comes with having a car. Many households (approx. 30 per cent) have more than one car, while 45 per cent have one car. Photo **D** shows some of the environmental problems this creates. Diagram **E** shows strategies designed to reduce the use of cars by encouraging cycling, making public transport more attractive, introducing **park-and-ride schemes** and congestion charging.

■ Air pollution from vehicles

■ Noise from heavy vehicles

■ Buildings discoloured

■ Impact on health – respiratory conditions, asthma

■ Unsightly

**D**  *Environmental problems resulting from traffic congestion*

### Study tip

Know the arguments in favour of urban development taking place on both brownfield and greenfield sites. Relate this to how an inner-city area has been changed as a direct result of a development strategy.

### Key terms

**Park-and-ride scheme**: a bus service run to key places from car parks on the edges of busy areas in order to reduce traffic flows and congestion in the city centre. Costs are low to encourage people to use the system – they are generally cheaper than fuel and car parking charges in the centre.

**Regeneration**: improving an area.

**Sustainable community**: community (offering housing, employment and recreation opportunities) that is broadly in balance with the environment and offers people a good quality of life.

**Quality of life**: how good a person's life is as measured by such things as quality of housing and environment, access to education, health care, how secure people feel and how happy they are with their lifestyle.

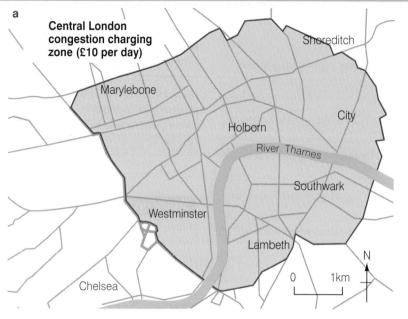

**a** Central London congestion charging zone (£10 per day)

**c**

**d**

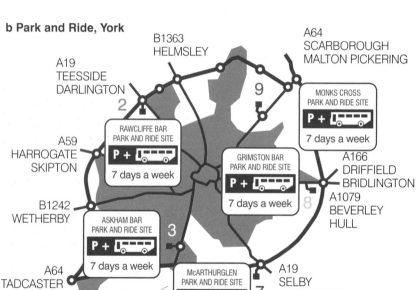

**b Park and Ride, York**

**e**

 **E**  *Attempts to solve the problem in London (top) and York (below)*

**Activity**

4  Study diagram **E**.

Produce a poster to show the good and bad points of the options available to solve the environmental impact of traffic.

# Traffic congestion and its management in Cambridge

## Why has traffic become a major problem?

The city of Cambridge with its world famous university has a population of 130,000. It has become a centre for hi-tech industry as well as being an administrative, tourism and retail centre. Because of its rapid growth in the 20th century, Cambridge has a congested road network. The M11 motorway from east London terminates to the north-west of the city where it joins the A14, a major freight route which connects the port of Felixstowe on the east coast with the Midlands. Many people commute to work in Cambridge and this results in heavy traffic congestion in the morning and early evening rush hours. In addition, the narrow streets of CBD (the historic core) result in slowing of traffic movement.

The problems include:

- increased air/noise pollution (especially in warm summer conditions) which contributes to health concerns
- parking shortages, resulting in congested streets
- increased travelling times
- possible delays for emergency services
- increased likelihood of traffic accidents.

There is no single answer to the problem but the City and County Council have introduced various schemes for traffic management. The main aims have been to remove through traffic, improve conditions for public transport, provide safer and convenient routes for cyclists, create better/safer environments for pedestrians and achieve an overall improvement in air quality. The schemes include:

- Pedestrianised streets. Since 1992 the streets in the historic core of the city have been closed to general passenger traffic, using a system of rising bollards.
- Cycle lanes. The city has the highest level of cycle use in the UK, with over 27 per cent of residents travelling to work by cycle. The city was chosen as a Cycling Town by the Department for Transport in 2008, with central government funding an expansion of cycling facilities in the city and its surrounding villages.
- City bypasses to the north and west of the city. These roads take more 'through-traffic' around the city and allow a more continuous, faster flow of traffic.
- Cambridgeshire Guided Busway (Photo **F**). This route, which opened in 2011, is the world's longest guided busway and passes through Cambridge. The route runs on normal roads for part of the journey, then a bus-only guided section along a former railway line south-westwards into Cambridge. Over 2.5 million passengers were carried in the first year of the service.

- Bus Lanes. This involves separate lanes for buses in rush hours and encourages people to use public transport more.
- Multi-storey car parks. These have provided more parking, although prices are kept fairly high to discourage people from taking cars into the city;
- Five Park and Ride schemes. Free car parking for over 4500 cars is provided at special bus terminals. The schemes reduce the number of vehicles entering the city centre. There are over 4 million park and ride journeys per year.
- Traffic calming in residential areas.
- Workplace parking charges and the encouragement of car sharing.
- Improving access and upgrading of Cambridge Railway Station. A second railway station for the city, at Cambridge Science Park, has also been approved by the government. Construction is due to begin in 2014 and completed in 2015.

**F**  *Cambridgeshire Guided Busway*

## Activities

5   Explain why traffic problems in Cambridge have become more serious in recent years.

6   Study the list of measures taken in Cambridge to reduce the problems of traffic congestion. Select three of these measures and explain why they might be considered sustainable.

7   Some experts say that time is running out for cities and their traffic. Explain this view.

## Issue 3: Economic decline

As older manufacturing industries have closed they have left empty, derelict buildings towards the centre of the city. Businesses can be put off by high land prices, lack of space, high crime and traffic congestion. Local authorities may have limited tax revenue so there is a lack of investment in the local area.

Modern industries need more space so tend to locate on the edge of the city. High unemployment in inner city areas (where the old industries were once located) leads to social problems.

Changes in shopping have also caused problems. City centre locations are no longer favoured. There has been a recent growth in out of town shopping centres as well as an expansion of online shopping, which has led to the decline of many CBDs. The cheaper land in these suburban locations also enables stores to operate on a larger scale and pass on the benefits of economies of scale to customers. Shop units in the CBD made vacant by large chain stores moving out, may stay empty as smaller independent stores cannot afford the high land rents.

### The central business district (CBD)

During the 1960s to the early 1980s, the CBD struggled to attract businesses. Out-of-town shopping areas and regional shopping centres became more favourable destinations as they offered pleasant shopping opportunities with ample parking. In contrast, city centres appeared busy and crowded. The air quality was poor, with the smell of diesel and lead concentrations in certain areas being a cause for concern.

However, there have been significant changes in CBDs and their image is now, once more, a positive one. They have become vibrant and pleasant places as a result of a number of initiatives.

G  The CBD in Leeds

### Case study

## Revitalising the CBD: Newcastle upon Tyne

Newcastle upon Tyne, in the North East of England, was once famous for its traditional industries of coal mining, ship building and heavy engineering. The city is now the commercial, educational and cultural focus for North East England. The Central Business District is in the centre of the city, bounded by Haymarket, Central Station and the Quayside areas.

The city centre shops declined from the 1970s to the 1990s partly because of competition from out of town retail parks such as the Metro Centre at Gateshead. There was a general lack of investment in the CBD which led to the dereliction of some buildings and a general deterioratrion in the shopping environment. In addition, the city centre suffered from crowding, poor air quality, a crime ridden image and poor parking availability.

Newcastle CBD has experienced large scale **regeneration** in recent years in an attempt to encourage people back to the city centre. It has built some landmark buildings, improved transport links, improved public safety, built new shopping areas and marketed itself better.

Two landmark projects are the Baltic Centre and The Sage. The Baltic Centre is a new art gallery built on the Tyne River in a former flour mill. The Sage is a multipurpose arts centre also built on the banks of the Tyne.

Transport and ease of movement has been improved by pedestrianising large areas of the CBD, building the Millennium bridge across the Tyne and improving the integration of Newcastle's rail, metro and bus system.

The main shopping centre Eldon Square has seen its parking facilities extended and its disabled access improved. The shopping area has been covered so that shoppers can avoid adverse weather. It has also been expanded and now accommodates 140 shops, including several new department stores and designer shops. A new bus station, replacing the old underground bus station, was officially opened in March 2007. A new shopping area, Eldon Gardens, has also been developed for smaller independent retailers. In nearby Grainger Town, grants were made available to retailers to take on derelict buildings.

Newcastle now also has more visual policing and better CCTV. Money has been spent on street furniture and paving, to ensure that the shopping environment is visually more attractive.

# Inner city regeneration: Manchester

**Sustainable communities** allow people to live in an area where there is housing of an appropriate standard to offer reasonable **quality of life**, with access to a job, education and health care. This initiative began in 2003 and one area affected by it is an area of east Manchester, formerly known as Cardroom and now renamed New Islington Millennium Village. As diagram **H** shows, it seeks to provide for an appropriate quality of life in inner-city Manchester in the 21st century.

## Sustainable communities example: New Islington Millennium Village

### What's coming to New Islington?

**New homes**
- 66 houses, 200 ground-floor apartments
- 500 two- and three-storey apartments
- 600 1- and 2-bed apartments
- 34 urban barns
- workshops
- refurbishment of Ancoats hospital and Stubbs Mill
- new office space

**Waterways**
- 3,000 metres of canalside
- 12 bridges
- 3 giant canopies
- 50 moorings for narrow boats and canalside facilities

**Urban amenities**
- 10 new shops
- 2 pubs, 2 restaurants, cafés and bars
- Metrolink stop in 10 minutes' walking distance
- New bus lines and bus stops
- 200 on-street and 1,200 underground car-parking spaces
- a safe Old Mill Street

**Parks and gardens**
- 300 new trees
- 2 garden islands, an orchard, a beach
- play areas and climbing rocks
- secured courtyard gardens
- private gardens and patios

**Community facilities**
- a primary school and play areas
- a health centre with 8 GPs
- 2 workshops
- a crèche
- an angling club and a village hall
- a football pitch

**Sustainability Agenda**
- boreholes will provide up to 25 litres per second of naturally filtered water
- central heat and power to generate 600 kW electrical energy and 1,000 kW thermal energy
- recycling collection points that allow occupants to recycle 50 per cent of domestic waste

 *Living in New Islington, Manchester*

## Activities

8    Study photo **G** opposite.

a    Use evidence from the photo to suggest how the CBD has been revitalised (improved).

b    Can you suggest any further improvements that would make the CBD an attractive area to visit?

9    What are the main differences in land use and function between the CBD and inner city?

10    Discuss the effects of out-of-town shopping centres on CBD shopping areas.

11    How can city centre shopping areas compete with out-of-town centres?

12    How has the shopping experience in Newcastle city centre been improved in recent years?

13    Complete a fact file about the New Islington Millenium Village project including the following information:
- Location
- Dates
- Aims
- Improvements to housing
- Environmental improvements
- Community improvements
- Sustainability

# 6.4   What are the issues for people living in squatter settlements in poorer parts of the world?

The speed of urbanisation in many poorer areas of the world results in **squatter settlements** being built and the evolution of an **informal sector** of the economy. The pace of rural–urban migration is too fast to allow the time needed to build proper houses. New arrivals to the city find unoccupied areas of land and building materials and begin to build their own makeshift shelters. As there are few official jobs available, people create their own employment by selling items, making and repairing things and becoming couriers, cleaners and gardeners.

**In this section you will learn**

why squatter settlements have developed, their characteristics and effects on people's lives

different strategies to try to improve squatter settlements and evaluate them

how to apply general concepts relating to squatter settlements to a case study.

**Key terms**

**Squatter settlements**: areas of cities (usually on the outskirts) that are built by people from any materials they can find on land that does not belong to them. They have different names in different parts of the world (e.g. *favela* in Brazil) and are often known as shanty towns.

**Informal sector**: that part of the economy where jobs are created by people to try to get an income (e.g. taking in washing, mending bicycles) and which are not recognised in official figures.

a Roçinha, the largest favela in South America

b Close-up of Roçinha

c Mumbai

**A**   *Typical shanty towns*

## Living in squatter settlements

No preparation for the building of these areas is done. The houses are not provided with basic infrastructure such as sanitation, piped water, electricity and road access. The photos in **A** (page 130) include views of Roçinha, a *favela* in Rio de Janeiro, which has a population of over 100,000. Close-up views give a clearer impression of conditions in the *favela*. The houses are made of any materials available nearby – corrugated iron, pieces of board – haphazardly assembled to provide a basic shelter. There is a simple layout that may have a living area separate from a sleeping area.

Parents and large families inhabit a small shack in an area that is very overcrowded. There are no toilets. Water must be collected from a nearby source – often at a cost – and carried back. Rubbish is not collected and the area quickly degenerates into a place of filth and disease. The inhabitants tend to have poorly paid jobs where the income is unreliable.

The quality of life for many people is poor. Houses are densely packed together with few open spaces and green areas or trees. Poor environmental conditions lead to disease and, with many people living in poverty, crime is a major issue.

**Key terms**

**Self-help**: sometimes known as assisted self-help (ASH), this is where local authorities help the squatter settlement residents to improve their homes by offering finance in the form of loans or grants and often installing water, sanitation, etc.

**Site and service**: occur where land is divided into individual plots and water, sanitation, electricity and basic track layout are supplied before any building by residents begins.

## Strategies to improve living conditions

Improvement by residents involves the residents seeking to 'do up' their original shelters (photo **B**). This means replacing flimsy, temporary materials with more permanent brick and concrete, catching rainwater in a tank on the roof, and obtaining an electricity supply (often illegally by tapping into a nearby source). Such improvements are slow and not all the problems can be solved.

**Self-help** occurs where local authorities support the residents of the squatter settlements in improving their homes. This involves the improvements outlined above, but it is more organised. There is cooperation between residents to work together and remove rubbish. There is also cooperation from the local authority, which offers grants, cheap loans and possibly materials to encourage improvements to take place. Standpipes are likely to be provided for access to water supply and sanitation. Collectively, the residents, with help from the local authority, may begin to build health centres and schools. Legal ownership of the land is granted to encourage improvements to take place, marking an acceptance of the housing.

**B**   *DIY improvements in Roçinha*

**Site and service** schemes are a more formal way of helping squatter settlement residents. Land is identified for the scheme. The infrastructure is laid in advance of settlement, so that water, sanitation and electricity are properly supplied to individually marked plots. People then build their homes using whatever materials they can afford at the time. They can add to and improve the structure if finances allow later.

Local authority schemes can take a number of different forms. There may be large-scale improvements made to some squatter settlements or new towns may be constructed. In Cairo, new settlements such as 10th of Ramadan City were built to reduce pressure on the city. High-rise blocks of flats were built, together with shops, a primary school and a mosque. Industries were also planned to provide jobs for the new inhabitants.

# Kibera, a squatter settlement in Nairobi, Kenya

Map **C** shows the situation of Kibera in the capital city of Nairobi in Kenya. Some 60 per cent of Nairobi's inhabitants live in slums, over half of them in Kibera. Specific facts about Kibera are uncertain. It is believed that between 800,000 and 1 million people live in the shanty town, in an area of 255 ha. This gives extremely high densities, with people only having 1 m² of floor space each. Over 100,000 children are believed to be orphans as a result of the high incidence of HIV/Aids. Photos **D** and **E** show the squatter settlement. The homes are made of mud, plastered-over boards, wood or corrugated iron sheeting. The paths between the houses are irregular, narrow and often have a ditch running down the middle that has sewage in it. Rubbish litters the area as it is not collected. The area smells of the charcoal used to provide fuel and of human waste. A standpipe may supply water for up to 40 families; private operators run hosepipes into the area and charge double the going rate for water. Crime is rife and vigilante groups offer security – at a price. Police are reluctant to enter. However, homes are kept clean and the residents welcome visitors.

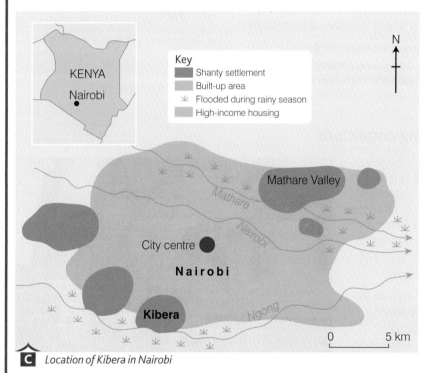

**C**  *Location of Kibera in Nairobi*

Key
- Shanty settlement
- Built-up area
- Flooded during rainy season
- High-income housing

KENYA
Nairobi

N

Mathare Valley
Mathare
Nairobi
City centre
**Nairobi**
**Kibera**
Ngong

0     5 km

**D**  *General view of Kibera*

## Finding solutions

There are signs that things are improving.

- Practical Action, a British charity, has been responsible for developing low-cost roofing tiles. Made from sand and clay, and using natural fibres and lime, building blocks can be made as a cheaper alternative to concrete. These allow self-help schemes to progress.

- The United Nation's Human Settlement Programme (UN-Habitat) has provided affordable electricity to some parts at 300 Kenyan shillings (about £2.25) per shack.

- There are two mains water pipes – one provided by the council and the other by the World Bank – at a cost of 3 Kenyan shillings (about 2 pence) for 20 litres. Improving sanitation is more difficult and progress is slow.

- Medical facilities are provided by charities.

- Gap-year students are encouraged to go to Kibera to oversee the spending and to help coordinate efforts.

- In a 15-year project that began in 2003, there are plans to re-house thousands of residents of Kibera. This is a joint venture between the Kenyan government and UN-Habitat. In its first year, 770 families were rehoused in new blocks of flats with running water, toilets, showers and electricity.

**E**   *Close-up view of Kibera*

**Activities**

**3**   Study map **C** and photos **D** and **E**. Produce a fact file on Kibera by completing the following tasks:

a   Draw a labelled sketch map to show the location of Kibera in Nairobi.

b   Write a summary of the key points about the Kibera squatter settlement.

c   Draw a labelled sketch from one of the photos to show the characteristics of the squatter settlement and the living conditions there.

d   Summarise the attempts that have been made to improve Kibera.

e   In your opinion, is Kibera a slum of hope (where things are getting better) or a slum of despair (where there is no improvement and no hope for the future)? Justify your views.

---

**Did you know** ??????

In 2009, about 35 per cent of Brazil's population lived in slums, about 50 million people. In Kenya, 79.2 per cent of the population lived in slums – 41,595,000 people.

**Study tip**

Look at the characteristics of squatter settlements and what improvement schemes are taking place. Look for examples in specific squatter settlements.

**Did you know** ??????

A special report by the BBC in 2002 described Kibera as 'Six hundred acres of mud and filth, with a brown stream dribbling down the middle. You won't find it on a tourist or any other map. It's a squatter's camp – an illegal, forgotten city – and at least one third of Nairobi lives here'.

**∞ links**

Investigate Kibera at **www.bbc.co.uk** and **www.mojamoja.org**

# 6.5   What are the problems of rapid urbanisation in poorer parts of the world?

## ▉ Examples of problems

### Environmental disasters

During the early hours of 3 December 1984, the world's worst industrial accident unfolded in the Indian city of Bhopal. Poisonous gas escaped from a chemical plant and killed at least 3,000 people (photo **A**). Around 50,000 suffered permanent disabilities and more died later. This is one example of how rapid urbanisation and **industrialisation** can lead to environmental problems in poorer parts of the world. Expanding cities lead to problems of air and water pollution and **disposal of waste**, including toxic waste from plants like the one at Bhopal. Non-existent or poor regulations and a lack of planning for an environmental emergency make problems worse.

### Electronic waste

Creation of electronic waste is another major problem in a rapidly industrialising country like India. The country imports more than 4.5 million new computers a year, plus many second-hand ones with shorter lifespans. Computer waste is known as electronic waste or e-waste. In the cities, India's poor scrape a living by breaking down PCs and monitors (photo **B**). They boil, crush or burn parts in order to extract valuable materials like gold or platinum. But what they do not realise is that the toxic chemicals inside like cadmium and lead can pose serious health risks. India's hospitals are starting to see patients with ten times the expected level of lead in their blood. Dumping and unsupervised recycling of e-waste is literally leading to a brain drain.

### Water pollution

The Ganges River contains untreated sewage, cremated remains, chemicals and disease-causing microbes. Cows wade in the river. People wash their laundry in it and drink from it (see Photo **D**).

### Waste and pollution

In Shanghai, the construction boom is creating 30,000 tonnes of waste per day. Industry is responsible for 70 per cent of the country's carbon dioxide emissions. Some 73 per cent of electricity is produced by coal-fired power stations. These factors are responsible for 400,000 deaths annually.

**Key terms**

**Industrialisation**: a process, usually associated with the development of an economy, where an increasing proportion of people work in industry.

**Disposal of waste**: safely getting rid of unwanted items such as solid waste.

**A**   *Deaths caused by the industrial accident in Bhopal*

**B**   *Breaking down PCs and monitors in India*

## Reducing the problems

In order to seek to reduce the environmental problems resulting from rapid urbanisation and industrialisation, there need to be guidelines to indicate what is allowed and what is not. Limits must be monitored and enforced to ensure that industries, for example, do not exceed the stated limits.

### Waste disposal

Waste provides a resource and a means of making a living for many shanty dwellers in poor countries.

- In São Paulo, Brazil, two huge incinerators burn 7,500 tonnes of waste a day, resulting in a problem caused by a management strategy. There were only two **landfill** sites in 1990. Children and adults alike scavenge and extract materials and then reuse or resell them. For example, car tyres may be made into sandals and food waste is fed to animals or used as a fertiliser on vegetable plots.

- In Shanghai, China, an effective solid waste disposal unit has been installed in most houses and the waste is used as a fertiliser in surrounding rural areas.

- Toxic waste and its safe disposal is a key issue in areas where the manufacturing industry is increasing. In the aftermath of the Bhopal accident in 1984, the site was covered in toxic waste. This could not be disposed of safely in India. This meant that the waste was packed up and sent to the USA.

- Large companies need to take responsibility for safely disposing of electrical goods in areas such as Bangalore in India, where there are many call centres. The large amount of e-waste in Bangalore is covered by one enforcement order, which is inadequate. There are not enough people employed to make sure the law is obeyed.

- **Recycling** plants are becoming more common in cities throughout the world as an attempt to reduce the need to dispose of waste in landfill sites (photo **C**).

## Air pollution

**Air pollution** is an issue in many industrial cities. There is a need to encourage the use of new technologies that can reduce emissions of sulphur dioxide and nitrogen oxide. Switching to cleaner, alternative sources of energy is an option. However, given the plentiful supplies of coal in countries such as China (where 80 per cent of electricity is from this source) and India, this may need the introduction of a carbon tax to induce a change. In Shanghai, China, industries use low sulphur coal to try to reduce pollution. Limits need to be set and enforced on emissions, and companies, including **transnational corporations (TNCs)**, must be monitored to ensure that emissions of carbon dioxide and sulphur dioxide are reduced. Transport also needs to be considered and strategies such as allowing odd-numbered cars into Mexico City on one day and even-numbered on another day can reduce traffic in towns. Other strategies include improving public transport, limiting the number of cars and introducing congestion charging to discourage car owners from entering city centres.

## Water pollution

As with air pollution, limits relating to **water pollution** need to be identified and enforced if quality is to be improved. In 1986, the Ganga Action Plan sought to introduce water treatment works on the River Ganges in India, which it did successfully. However, the increasing population was not taken into account and water quality has since deteriorated. Such attempts have been replicated in other countries. In Shanghai, the Huangpu and Suzhou rivers have been the target for improving water quality. A World Bank loan of $200 million was granted to this cause in 2002.

## links

For information on Union Carbide, the Bhopal disaster and toxic e-waste, go to **www.bbc.co.uk**.

For reducing air pollution in Asia and the Pacific, visit **www.unep.org**.

Look into the dumping of toxic waste in poor countries such as Ivory Coast at **www.npr.org** and research more about e-waste at **www.bbc.co.uk**.

**C**  *A recycling centre in Beijing, China*

**D**  *The River Ganges, India*

## Activities

1
a   Describe the sources of pollution in the rivers Ganges and Huangpu.

b   Explain why they present a hazard to health.

c   Produce a fact file to summarise the events at Bhopal on 3 December 1984.

d   What is your view of the events of that night? Give reasons for your opinion.

e   Why do many poorer countries use coal as a source of energy?

f   Describe how some people living in cities in India make a living from discarded computers.

g   Explain the dangers of making a living in this way.

2   Study the text on the strategies to reduce the environmental effects of rapid urbanisation.

Produce a diagram to summarise the strategies adopted to manage the environmental problems in poorer parts of the world. Use case studies and illustrations as part of your work.

# 6.6 How can urban living be sustainable?

A **sustainable city** has certain characteristics that relate to its long-term future. It enjoys a clean environment, has a sound economic base with plenty of jobs and has a strong sense of community with local people involved in decision making. Sustainable cities make use of public transport, manage waste efficiently and create green spaces and gardens. Sustainable cities are cities of the future! Photos **A** and **B** show two contrasting urban scenes in Los Angeles and Belfast.

> **In this section you will learn**
>
> attempts made to ensure that city life is environmentally and socially sustainable
>
> the characteristics of a sustainable city.

**A** *Los Angeles: non-sustainable urban living?*

> **Key term**
>
> **Sustainable city**: an urban area where residents have a way of life that will last a long time. The environment is not damaged and the economic and social fabric are able to stand the test of time.

**B** *Belfast: sustainable urban living?*

> **Activity**
>
> 1 Study photos **A** and **B**.
>
> a Describe the characteristics of the urban scenes from the photos.
>
> b In which of the photos do you think life is sustainable? Explain reasons for your choice.

## Seeking environmental sustainability

### Conserving the historic environment

The Liverpool Maritime Mercantile City provides an example of conserving an area of previous industrial use and historic commercial and cultural areas. The Liverpool waterfront and areas associated with its development were designated a World Heritage Site in 2004. The award recognised the importance of the area as a port and associated buildings of global significance during the heyday of the British Empire in the 18th and 19th centuries. Many of the buildings are architecturally as they were then, although their function has changed. The photos in **C** show some of the many faces of the sites that gained recognition, conserving an environment so rich in history and heritage.

**Key term**

**Sustainability**: development that looks after future resources and considers the needs of future generations.

a Liver Building

b Cunard Building

c Albert Dock Warehouse

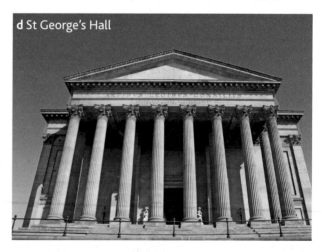

d St George's Hall

**C** *Liverpool maritime mercantile city*

**Activity**

2   Study the photos in **C**. Produce a leaflet for Liverpool City Council or another city near where you live. Produce an illustrated report to show how the city has used and conserved its industrial heritage in a sustainable way. Use the internet to help you with your research.

## Conserving the natural environment

**E**   *Advantages of building on a brownfield site*

- Provides an economic use of derelict land
- Provides jobs
- Foundations often in place for new construction
- Improves environment
- No need to develop greenfield sites
- Prevents **urban sprawl**

**D**   *Green belts in England*

The natural environment can be conserved by restricting development on the edge of a town or city. **Green belts** exist around many large towns in England (map **D**). They were established to protect the surrounding countryside from development. This often provides recreational open space for urban residents. Limiting available sites on the edge of the city means that alternative locations for development must be offered if growth is to continue. This means that building on brownfield sites is encouraged. There are several advantages of building on brownfield sites (diagram **E**).

## Reducing and safely disposing of waste

By 2008, the UK was producing 400 million tonnes of waste each year – enough to fill the Royal Albert Hall in London *every hour*. Much of this was from mining and quarrying, but 30 million tonnes was from households, many of them in cities. There is a need to reduce the amount of waste produced. A great deal of domestic waste can be composted (20 per cent of household waste is garden waste and 17 per cent kitchen waste) or recycled (18 per cent is paper/card). In 2011, around 40 per cent of UK household waste was recycled. That is a big improvement from 2001, when only 11 per cent was recycled. But the UK still sends more waste to landfill than most European countries.

**Key terms**

**Urban sprawl**: the spreading of urban areas into the surrounding rural/rural–urban fringe areas.

**Green belt**: land on the edge of the built-up area, where restrictions are placed on building to prevent the expansion of towns and cities and to protect the natural environment.

**Did you know** ??????

In the UK we produce 275 kg of waste per person every year.

It is important to reduce waste so that fewer plastic bags are used. Consumer pressure could reduce packaging in general – do apples need to come in plastic bags? Do red peppers need individual packaging? Packaging can be made so that it can be returned and reused, such as milk bottles and 'bag for life' carrier bags.

At present, over 80 per cent of household waste is disposed of in landfill sites (photo **F**). However, in the future, there will be a shortage of suitable sites. Incineration (burning) is another option, accounting for 8 per cent of waste disposal. Although incineration can produce energy (photo **G**), it is a controversial strategy. For example, a new facility planned for Kings Lynn is unpopular with local people due to fears about pollution and increased traffic.

## Providing adequate open spaces

The presence of official green belts or areas where local authorities choose to restrict buildings around cities offers open space for recreation purposes. In addition, many areas within cities have designated areas of open space in the form of parks, playing fields and individual gardens. Map **H** shows the distribution of open space in Greater London and some of the types of open space available.

**F**  *A landfill site in Liverpool*

**G**  *Incineration: Sheffield energy recovery facility*

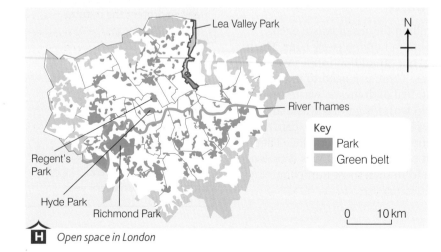

**H**  *Open space in London*

Lea Valley Park

River Thames

Regent's Park

Hyde Park

Richmond Park

N

Key
■ Park
■ Green belt

0    10 km

**Did you know** ??????

Before many retailers introduced charges for single-use carrier bags and offered 'bags for life', UK shoppers used 10.7 billion new single-use carrier bags a year. This figure had dropped to 6.4 billion by the end of 2010, a reduction of 40% but has since increased by 5% during 2011.

## Activities

**3** Study map **D** and diagram **E**. Imagine you are a town planner on a visit to a local school where you have to explain how green belt policy and the use of brownfield sites can help in conserving the natural environment. Design a presentation using Powerpoint to outline the advantages of green belts and using brownfield sites for development.

**4** Work in pairs. Produce a radio advert for your local council encouraging residents to 'reduce, reuse and recycle'. Make the advert clear and informative, and try to introduce a catchy slogan to promote a more responsible attitude to waste.

**5** Study photos **F** and **G**. Summarise the advantages and disadvantages of:

- landfill
- incineration

as a means of waste disposal.

**6** a What materials are collected by the council from your home?

 b What other items do you recycle at home?

 c How do you reuse items at home?

 d Describe how you try to reduce waste at home.

 e What efforts are made at school to reduce, reuse and recycle?

 f How effective do you think efforts are at home and at school? In what ways could they be improved?

**7** Study map **H**.

 a Describe the distribution of the different types of open space in Greater London.

 b London was designed to have many central parks. How do you think these parks are used by people?

 c Do you think cities should have open spaces such as parks? Justify your answer.

## Involving local people

If people have ownership of ideas and feel involved and in control of their own destiny, they are much more likely to respond positively and care for the building and environment in which they live. Consulting people at planning stages – before decisions are made – is essential. Planners increasingly survey opinions before putting forward plans and consult after they have been produced. Residents form associations to give them a stronger collective voice (photo **I**).

## Providing an efficient public transport system

The volume of cars as a means of private transport is a problem and a barrier to a city being sustainable. London has sought to make parts of the city unattractive to drivers by congestion charging. However, an alternative form of transport needs to be available in the form of fast and efficient public transport. This means a focus on the Underground and improvements in buses and rail links. The 2012 Olympics added impetus to improvements. The Tube was upgraded, not just the lines, but also the trains and stations. London overground links have been extended to form a complete circuit around London – the railway equivalent to the M25. Buses have also been improved to run more frequently to reduce overcrowding.

All buses now have CCTV to increase feelings of security and bus shelters have been added at more bus stops. The extension of bus lanes has led to quicker journeys and cash fares have been frozen. Schemes such as the Oyster card, which allows for the advanced purchase of up to £90 worth of journeys on a swipe card, offer journeys at reduced rates.

**I** *Resident's meeting*

### Did you know ??????

Between three and a half and four million journeys are made on the Underground (Tube) in London every weekday. The Tube is the oldest of its kind in the world – some parts dating back to the 1860s.

Case study

# Sustainable urban living in Curitiba

Curitiba is the capital city of the Brazilian state of Parana. It is the seventh largest city with a population of 1.8million. The city is seen as a role model for planning and sustainability in cities worldwide. In 1968, the Curitiba Master Plan was adopted to controlurban sprawl, reduce traffic in the city centre, develop public transport and preserve the historic sector. The emphasis has been on ensuring an appropriate quality of life for the residents of Curitiba with concern for the environment and the need to leave a suitable area for future generations living in the city.

There is now a network of 28 riverside parks creating almost 100 miles of city trails. Lakes have been created within these parks that fill and flood the surrounding parkland in periods of heavy rain reducing the risk of flooding in the city itself.

The 'green exchange' programme involves low-income families in shanty towns exchanging rubbish for bus tickets and food. 70 per cent of the city's rubbish recycled by its residents.

## The Bus Rapid Transport (BRT) System

Curitiba was the first Brazilian city to have dedicated bus lanes. The BRT system has four elements:

- direct line buses operate from key pick-up points
- speedy buses operate on the five main routes into the city and have linked stops
- inter-district buses join up districts without crossing the city centre
- feeder mini-buses pick people up from residential areas.

## Housing in Curitiba

In Curitiba, COHAB, the city's public housing programme, believes that residents should have 'homes – not just shelters'. They have introduced a housing policy that will provide 50,000 homes for the urban poor.

**J** *The innovative bus system in Curitiba*

## Activities

**8** Study the information in this case study and use the internet to find out more about Curitiba.

a Describe how Curitiba's bus system works and assess how successful it is.

b Explain why it was a good idea to preserve part of the old area of Curitiba.

c In Curitiba, 'planners come up with creative and inexpensive ways to go about solving universal problems for cities'. Provide evidence to support this statement.

**9** Working in pairs, summarise the characteristics of a sustainable city.

# 7 Globalisation in the contemporary world

## 7.1 What is globalisation?

Have you heard the phrase 'The world is shrinking'? It's often used to describe the way that technology such as satellites, the internet and high speed travel have reduced the 'distance' between places. Everywhere seems closer together and more interconnected. Consider all the world brands in the High Street. This increasing connectivity and **interdependence** in the world is what is meant by **globalisation**.

This has been made possible by the relaxation of laws allowing foreign investment in countries (which encouraged the rise of transnational corporations, or TNCs), the increased provision and speed of international transport and developments in communication with the use of fax, telephone and e-mail.

Nike is a TNC that manufactures footwear and clothing. It has 148 contract plants in China, 55 in Vietnam, 41 in Thailand, 15 in South Korea and others elsewhere in Asia. In addition, it has factories in South America, Australia, Canada, Italy, Turkey and the USA. Africa is the only inhabited continent not represented.

> **In this section you will learn**
>
> what is meant by the term 'globalisation'
>
> about the increased interdependence and interrelationships that result from greater connectivity between the different countries.

> **Key terms**
>
> **Globalisation**: the increasing links between different countries throughout the world and the greater interdependence that results from this.
>
> **Interdependence**: the relationship between two or more countries, usually in terms of trade.

> **Study tip**
>
> Ensure that you explain the meanings of the terms 'globalisation', 'interdependence' and 'transnational corporation' (defined on page 147).

**A**   *Nike shoes on sale in China*

## ■ The production of a Wimbledon tennis ball

Slazenger has supplied the All England Club since 1902. Dunlop Slazenger is responsible for making the 48,000 balls that supply the showpiece tournament in June each year. From the 1940s to 2002, the tennis balls were made at Barnsley, south Yorkshire. However, in an attempt to boost profits by cutting labour costs, production shifted to Bataan in the Philippines in 2002.

Table **B** shows the ingredients for making the tennis balls and their origin. Many of the components are used to vulcanise the rubber so that the balls have the correct amount of bounce. They are all transported to Bataan for manufacture, before being despatched to London.

> **Did you know** ???????
>
> 300 million tennis balls are made each year, 90 per cent of them in South-east Asia. Wimbledon tennis balls are recycled to provide homes for the harvest mouse, which is a threatened species.

**B** *The manufacture of a Wimbledon tennis ball*

| Ingredient | Origin | Destination |
| --- | --- | --- |
| Wool | New Zealand | UK |
| Cloth (made from wool) | UK | Philippines |
| Dyes | UK | Philippines |
| Silica | Greece | Philippines |
| Zinc oxide | Thailand | Philippines |
| Rubber | Malaysia | Philippines |
| Tins | Indonesia | Philippines |
| Clay | USA | Philippines |
| Magnesium carbonate | Japan | Philippines |
| Sulphur | South Korea | Philippines |
| Tennis balls (product) | Philippines | UK |

**C** *A Wimbledon tennis match*

## Activities

**1** a Write down a definition of 'globalisation' and give one example to illustrate it.

b Compare your answer with a partner's and agree on a definition.

c Now work together to make a list of international companies that have shops or businesses in your local town or city centre.

**2** Study table **B**.

a On a world political map outline, draw flow lines to show the movement of materials to the Philippines and then the completed tennis balls to the UK.

b Describe how your map illustrates the concept of 'globalisation'.

c Why might it be said that the production of the cloth to cover the tennis balls is the best example of the globalisation process?

⚭ **links**

Find out more about globalisation by typing the word into the search box at **www.bbc.co.uk**.

# 7.2   What are the factors influencing globalisation?

## The influence of developments in ICT and transport

Improvements in transport and communications have been largely responsible for the spread of manufacturing and services worldwide. Look at Graph **A**. Notice how transport and communication costs have fallen over the years.

Much of the advances made in communication are the result of developments in satellites. A satellite is an object that revolves around the earth following a particular path or orbit. It is usually built for a specific purpose, for example, weather or communications satellites (Photo **B**) . Consider how much the media depends on satellite for television (i.e. Sky) and for transmitting sports fixtures 'live' around the world.

> ### In this section you will learn
> how developments in ICT have encouraged globalisation
>
> how specialist, localised industrial areas have developed with global connections, e.g. Motorsport Valley®
>
> why call centres have developed abroad.

1930 costs = 100    1970 costs = 100

Legend:
- Transatlantic phone call [1]
- Sea freight [2]
- Air transport [3]
- Satellite charges

Original source: HM Treasury

[1] Cost of three-minute telephone call from New York to London
[2] Average ocean freight and port charges per short ton of import and export cargo
[3] Average air transport revenue per passenger mile

**A**   *Falling transport and communication costs*

The development of submarine cables has been important in allowing global operations for both manufacturing and service industry. These underwater cables link western Europe with SE Asia, the Middle East and Australia, providing high speed transmission of data.

Developments in ICT have allowed immediate access to people all over the world and fostered developments in small areas, knowing that communication across the world is possible.

> ### Did you know ??????
> Submarine cables can be accidentally damaged. On 26 December 2006, an earthquake damaged SEA-ME-WE3 cables off Taiwan and, on 30 January 2008, a ship's anchor damaged the SEA-ME-WE4 off Egypt.

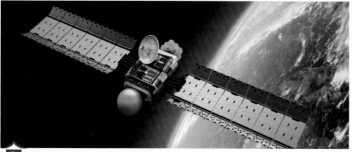

**B**   *Satellites are used for high-speed communications*

## ■ The development of call centres abroad

A call centre is an office where people respond to telephone enquiries, often relating to banking, communications (internet, telephones, etc.) and the media. Call centres are big business and in the UK they employ 400,000 workers, often in small towns such as Harrogate, Carlisle, Gateshead and Warrington. Banks and other finance companies such as insurance were among the first to develop centralised call centres and the first to look abroad. Household names such as ASDA, Tesco, BA, BT, Barclays, Lloyds TSB, HSBC and Virgin Media have all set up call centres in India. It is the big cities that house these, such as Mumbai, Delhi, Hyderabad and Bangalore (map **C**). Other important destinations for call centres abroad are South Africa and the Philippines.

About 10 per cent of the population (some 100 million people) speak English fluently

Of the 787 million living in towns, about 80 per cent are literate and 18 per cent of these are graduates

Operating costs are between 10 and 60 per cent lower than in the UK

Salaries are lower, e.g. £1,200 per year in contrast to £12,000 per year in the UK

Low staff turnover, working nine-hour shifts at times to fit in with origin country of company

Development of ICT allows fast and clear communication

**C** *Telephone call centres in India*

### Freedom of trade

Many barriers to trade have been removed. Some of this has been done by regional groupings of countries such as the EU or by the World Trade Organisation (WTO). This makes trade cheaper and therefore more attractive to business.

### Labour availability and skills

Countries such as India have lower labour costs (about a third of that of the UK) and also high skill levels. Labour intensive industries such as clothing can take advantage of cheaper labour costs and reduced legal restrictions in LEDCs.

### Growing global markets

More and more people living in major world cities have enough wealth to be significant consumers of goods and services. In 2011, China already had an estimated 50 million wealthy consumers, and it is predicted to become the world's largest market for consumer goods by 2015. In the emerging markets of South East Asia and Latin America there has also been considerable growth in incomes. The arrival of global satellite television has exposed consumers to global advertising. Consumers are more aware of what is available in other countries, and are keen to purchase new items.

### ∞ links

You can find out more about submarine cables at **www.telegeography.com**.

Visit **www.call-centres.com** for more on call centres.

**Activities**

1 Study graph **A** on page 145.

a What is the significance of the 100 value on the y-axis?

b What happened to the cost of a transatlantic phone call between 1930 and 1950?

c State two changes in communication costs between 1930 and 2000.

d Give evidence from the graph to support the changes described in c.

2 Study map **C**. Work in pairs for this activity. Imagine you are director of customer relations for a call centre in India. You are being interviewed about your reasons for choosing to locate abroad for national television news. Present the questions you were asked and your answers to them.

# 7.3    What are the advantages and disadvantages of TNCs?

**Transnational corporations (TNCs)** are large, wealthy international organisations. They are companies that have their headquarters in one country, but often have many other branches spread across much of the world. As a result of improvements in transport and communications, TNCs have grown steadily over the last 30 years.

Most TNCs have their headquarters in richer areas of the world, especially the USA, UK, France, Germany and Japan. Research and development is usually centred here. Production often occurs in poorer areas where labour costs are lower, laws more lenient or where governments want to seek investment. Production also occurs in richer areas where benefits include a skilled workforce. Table **A** shows the biggest TNCs according to their value.

**In this section you will learn**

what advantages and disadvantages result from the presence of TNCs in countries

the characteristics of Toyota as a case study of a TNC.

**A**    *The top ten non-financial TNCs, 2008*

| Rank | TNC | Headquarters | Product |
|------|-----|-------------|---------|
| 1 | General Electric | USA | Electrical/electronic equipment |
| 2 | British Petroleum | UK | Petroleum |
| 3 | Toyota | Japan | Motor vehicles |
| 4 | Royal Dutch/Shell | UK/Netherlands | Petroleum |
| 5 | ExxonMobil | USA | Petroleum |
| 6 | Ford | USA | Motor vehicles |
| 7 | Vodafone | UK | Telecommunications |
| 8 | Total | France | Petroleum |
| 9 | Electricité de France | France | Electricity/gas/water |
| 10 | Wal-Mart | USA | Retail |

**Key terms**

**Transnational corporation (TNC):** a corporation or enterprise that operates in more than one country.

**Multiplier effect:** where initial investment and jobs lead to a knock-on effect, creating more jobs and providing money to generate services.

**Leakage:** where profits made by the company are taken out of the country to the country of origin and so do not benefit the host country.

- **Advantages of TNCs**

TNCs offer many advantages to the countries where they set up branches. They provide jobs in factories making supplies and in services where products are sold. The additional income benefits local businesses, creating a **multiplier effect**. Training of the workforce leads to the development of skills. Often, the infrastructure is improved as better access, both within and between countries, and communications are needed.

- **Disadvantages of TNCs**

A significant disadvantage is that of **leakage**, as well as the fact that in some locations wages are very low and key jobs go to outsiders. If there are problems worldwide economically or within the company, the branch plants may be closed. In some areas, working conditions can be poor and the labour force is expected to work long hours. Health and safety may be an issue and pollution may be a problem in countries where there are less strict rules and regulations.

**Activity**

1    Study table **A**. Work with a partner.

a    Find out the identity of a further 10 large TNCs.

b    Produce a collage to display the name, logo and line of business in which they are involved. Try to do this in a limited amount of time – say, 20 minutes – to see how you can work under pressure.

# Toyota

Toyota began in Toyota, Aichi, Japan in 1937. Seventy years later, it had become the biggest producer of cars in the world, with profits of $11 billion in 2006. Map **B** shows (in red) the countries where Toyota plants are present and table **C** shows production in different regions in 1996 and 2005.

**B**   *The global location of Toyota*

| **C** | Toyota production by region, 1996 and 2009 (1 = 1,000 vehicles) | |
|---|---|---|
| **Region** | **1996** | **2009** |
| North America | 782.9 | 1,189.1 |
| Latin America | 3.2 | 181.5 |
| Europe | 150.3 | 507.3 |
| Africa | 85.0 | 102.8 |
| Asia | 257.0 | 1,501.4 |
| Oceania | 67.6 | 96.8 |
| Overseas total | 1,346.0 | 3,579.0 |
| Japan | 3,410.1 | 2,792.2 |
| Worldwide total | 4,756.1 | 6,371.3 |

Toyota began to develop overseas in the late 1950s in Brazil and decided to seek to develop production in the UK in the early 1990s. The reasons for this were:

- the strong tradition of car manufacturing in the UK
- the large domestic market
- excellent workforce and favourable working practices
- positive support from the UK government for inward investment
- familiarity with the English language in Japan, so
- communication and integration is much easier
- good industrial transport links with customers and suppliers in Europe.

In 1989 Toyota Manufacturing UK was established.

## Activities

**2**   Sort the following into advantages/disadvantages of TNCs to both host countries and countries of origin.

1. TNCs bring new capital, technology, skills and expertise to a country.
2. Employees in poor countries may have to work long hours in poor conditions.
3. TNCs are mobile and can leave a country as quickly as they arrived.
4. Outsourcing work to LEDCs may lead to job losses.
5. Cheaper imports may benefit consumers
6. Company profits may increase and shareholders benefit.
7. TNCs export profits, draining wealth from the country's economy.
8. Manufacturing industry declines and the local economy suffers.
9. Job creation in labour-intensive industries.
10. Infrastructure developments, such as roads, and railways benefit the whole country.
11. Most products are for export and subject to changes in price on the world market.
12. Exports increase, helping to raise living standards.
13. TNCs can damage the local environment and deplete resources.
14. Less rigid health, safety and pollution controls can adversely affect the local workforce and environment.
15. TNCs help to develop mineral wealth and improve energy production.
16. The profits made can be used for research and development in the country of origin.

**3**   Study map **B** and table **C** above.

a   With the help of an atlas, on an outline world map label 10 countries where Toyota is present. Make sure your labels reflect the global presence of Toyota.

b   On your map, draw located bars to show production in each region for 1996 and 2009.

c   Describe the pattern shown by your map in 2009.

d   Summarise the changes between 1996 and 2009.

e   Where and why do you think Toyota should locate future new branches?

# 7.4   How and why is manufacturing in different countries changing?

## ■ Changes in relative importance of world regions

Manufacturing industry has declined in importance in some regions, while it has become more significant in others. The changes in high-tech manufacturing are shown in graph **A**. In many of the richest areas of the world, manufacturing has declined. For example, in Britain the number of people employed in manufacturing fell from just over 6 million in 1981 to 2.49 million in 2010. This is the result of **de-industrialisation**, which has occurred because of increased mechanisation and the need for industry to be competitive.

> **In this section you will learn**
>
> how and why the relative importance of manufacturing is changing globally
>
> the emergence of China as the new economic giant.

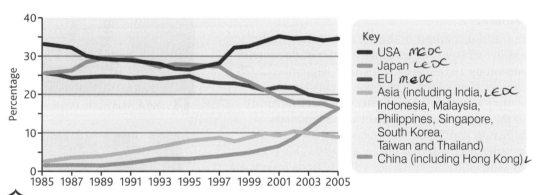

Key
- USA *meoc*
- Japan *ledc*
- EU *meoc*
- Asia (including India, *ledc* Indonesia, Malaysia, Philippines, Singapore, South Korea, Taiwan and Thailand)
- China (including Hong Kong)∠

**A**   *World share of high-tech manufacturing, 1985–2005*

## Reasons for changes

There are many reasons why some areas have experienced growth in manufacturing industry whereas others have suffered a decline.

### 1  Government legislation

This can take many forms, such as:

- setting up areas (**assisted areas/enterprise zones**) where conditions are favourable for new industry, for example lower taxes
- providing **advanced factories** of various sizes
- offering retraining and removal expenses
- ensuring educational reform is high on the list in areas such as the four Asian 'tigers'.

Some countries have a minimum wage. In the UK, this is currently £6.08 per hour for those aged 21 years and older (2012 figures), but in Sri Lanka garment workers are often paid a fraction of a much lower minimum wage.

An EU directive limits the maximum number of hours worked per week to 48 (except in the UK). The average South Korean works 2,390 hours a year in contrast to 1,652 in UK. This has real implications for production. In Sri Lanka, garment workers should not work after 10pm

> **Key terms**
>
> **De-industrialisation**: a process of decline in some types of industry over a long period of time. It results in fewer people being employed in this sector and falling production.
>
> **Assisted areas/enterprise zones**: areas that qualify for government help. Enterprise zones are on a smaller scale than assisted areas.
>
> **Advanced factories**: where buildings for production are built in the hope they will encourage businesses to buy or rent them.

due to International Labour Organization rules, but they are often forced to do so by their employers.

## 2  Health and safety regulations

Working conditions tend to vary globally. In the UK, adult employees working over six hours are entitled to a 20-minute break. Workers have the right to:

- know how to do their job safely and to be trained to do so
- know how to get first aid
- know what to do in an emergency
- be supplied with protective clothing.

Such regulations often do not exist in some poorer countries or are not enforced. Some workers – unable to travel home after a shift and arrive back on time for the next shift – sleep on the factory floor, although this is illegal.

## 3  Prohibition of strikes

In the 1970s, there was much unrest in the UK and various trade unions frequently called **strikes**. The so-called 'winter of discontent' began with strike action by 15,000 Ford workers in September 1978. There were many groups that joined in and power cuts became the norm. Such disruption had an adverse effect on manufacturing industry and led to the reduction of the power of trade unions. Companies such as Nissan and Toyota only came to the UK on the understanding that strike action would not be allowed. This is also true of many newly industrialising countries (NICs). Trade unions are allowed in Sri Lanka, but workers are likely to be threatened if they join one.

## 4  Tax incentives and tax-free zones

Tax incentives take a variety of forms, but all seek to offset costs. For example, One NorthEast (the development agency responsible for the north-east of England) offered job-creation grants, business rate or rent-free periods and help in preparing a business plan. Regional development agencies closed in 2012 with some of their functions now carried out by local enterprise partnerships.

Tax-free zones are areas where new businesses do not need to pay tax, e.g. parts of Dubai (photo **B**).

## ▪ China: the new industrial giant

China has been dubbed 'the new workshop of the world' (photo **C**), a phrase that was first used to describe Britain during the 19th century. China makes 60 per cent of the world's bicycles (photo **D**) and 72 per cent of the world's shoes. Between 2000 and 2006, cloth manufacture more than doubled and car production increased by more than six times. Mobile phone ownership increased nine times. The National Bureau of Statistics of China is predicting economic growth of at least 7 per cent to 2018. China took over 1st place in industrial output from the USA during 2012.

> **Key term**
>
> **Strikes:** periods of time when large numbers of employees refuse to work due to disagreements over pay or other grievances.

**B**  *Burj Al Arab Hotel, Dubai*

**C**  *In 2010 China became the world's largest exporter*

**D**  *60 per cent of the world's bicycles are made in China*

# Reasons for China's rapid growth

There are many reasons why China is emerging as the new economic giant.

## 1 Government legislation

In 1977, Deng Xiaoping sought to end China's isolation and stimulate Chinese industry. Foreign investment was encouraged but the government maintained overall control over the economy so that China would gain maximum benefit.

Between 1980 and 1994 special economic zones (SEZs) were set up. These paved the way for foreign investment by providing tax incentives to foreign companies.

## 2 The home market

The one-child policy, introduced in 1979, successfully reduced population growth and, as families become wealthier, consumer demand increased. Today China has a large and relatively rich urban population demanding electrical houshold goods, air conditioning, cars and computers. China's massive home market will continue to grow.

## 3 The Olympics factor

The 2008 Olympics were held in Beijing. This provided China with the perfect opportunity to showcase the nation. The opening ceremony, based on the theme 'One World, One Dream', was important in an attempt to convey China as a modern, open and friendly country.

## 4 Energy

Industrial development on a large scale demands large resources of energy. China currently generates two-thirds of its electricity at coal-fired power stations. Many new plants are being built. Hydroelectric power (HEP) accounted for 13.9 per cent of electricity in 2010. China produces more HEP than any other country in the world and is keen to develop new sources of energy. The Three Gorges Dam is the biggest in the world, generating 22,500 mW. Together with the development of navigation along the Yangtse, the dam has led to much development (see page 72).

## 5 Labour

Cheap labour is a key reason why the economy has been thriving. Wages are 95 per cent lower than in the USA.

# Advantages of industrial growth to China

The money China makes from exports can be used to buy more raw materials, or to invest more heavily in Chinese business and infrastructure.  It helps boost GDP (Gross Domestic Product) and therefore living standards. China's economic growth is on average 8 per cent per year and as a result poverty has declined dramatically over the past 30 years.

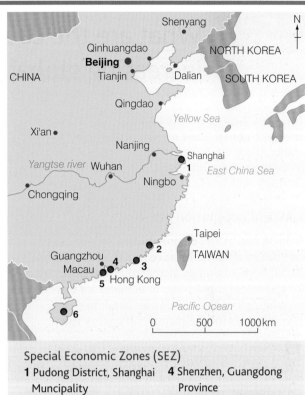

**Special Economic Zones (SEZ)**

**1** Pudong District, Shanghai Muncipality

**2** Xiamen, Fujian Province

**3** Shantou, Guangdong Province

**4** Shenzhen, Guangdong Province

**5** Zhuhai, Guangdong Province

**6** Hainan Province

**E**   *China's Special Economic Zones (SEZs)*

## Activities

**1**   Study graph **A** on page 149.

a   What percentage of high-tech manufacturing was produced in the EU in 1985 and 2005?

b   What percentage of high-tech manufacturing was produced in China in 1985 and 2005?

c   Using evidence from the graph, summarise the trends shown.

**2**   Study map **E** and the text on this page.

Produce an information poster (preferably A3 size) identifying the main factors responsible for China's recent economic growth. Use the Internet to find maps and photos to support text boxes. Choose a striking image for the centrepiece of your poster.

# 7.5   What are the causes and effects of increasing global demand for energy?

There are a number of reasons why there is an increase in demand for energy. Some of these reasons will be considered separately but, in reality, there are links between them. The increasing world population is getting richer and advances in technology increase the availability of products and create an insatiable thirst for energy.

## 1  World population growth

Table **A** and graph **B** show the overall global increase in population and the relative importance of different regions. Notice how the global population has increased dramatically since 1950 accounting for the recent rapid surge in demand for energy. The rapid rise in the populations of Asian countries combined with their economic growth also accounts for the increase in energy use.

**A**   *Overall global increase in population, 1750–2050*

| Year | 1750 | 1800 | 1850 | 1900 | 1950 | 1999 | 2050 |
|---|---|---|---|---|---|---|---|
| World population (billions) | 0.79 | 0.98 | 1.26 | 1.65 | 2.52 | 5.98 | 8.91 |

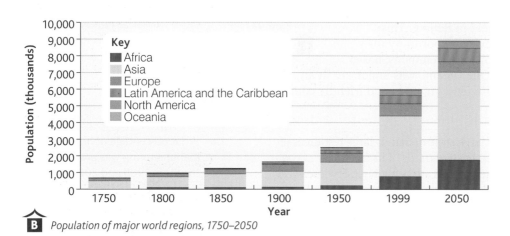

**B**   *Population of major world regions, 1750–2050*

## 2  Increased wealth

As people become wealthier they demand more products and consumer items, such as electrical devices. These require energy to make and to use.

Consider the following facts and the impact of these on demand for energy:

- The average wage in China has risen to 1750 yuan a month, four times higher than in 1995.
- Private car ownership in China increased from virtually zero in 1997 to 26 million in 2009.
- In the UK, the number of families not owning a car fell from 32 per cent to 27 per cent. In 2009 more households owned two cars than owned no cars at all.

**Did you know** ??????

World population is expected to stabilise at 10 billion. The UN predicts that this will be in 2200.

**Study tip**

Be clear as to how the social, economic and environmental impacts of increased energy use differ from each other.

## 3 Technological advances

Technological advances have supplied us with increasing amounts of energy and a wide variety of goods that we can purchase. The development of steam using coal led to large-scale production in the UK. It is the use of coal in power stations that has fuelled the Chinese economy. Modern technology allows the development of other sources of energy, such as nuclear power. Resources can now be exploited from deep below the sea and in very inhospitable regions, such as the Arctic. Research and development is big business. Companies strive to compete with each other and produce smaller mobile phones, faster laptops, in-house entertainment systems that offer best-quality pictures, most innovative games, etc. People see these and want to buy them. This requires energy to develop, make and run the consumer products that are so much a part of our lives today.

## ■ Environmental impacts

Some cities are shrouded in a haze that blocks out the sun and contains a dangerous mix of chemicals, including those from coal smoke and ozone. Photo **C** shows Beijing before the 2008 Olympic Games, when 1.3 million of the city's 3 million cars were taken off the road and 100 factories were closed. Poor air quality leads to asthma and other respiratory diseases.

Other environmental impacts include:

- on land, where spoil heaps have built up adjacent to coal mines when unneeded material has been dumped

- on water, where the transportation of oil has led to major pollution incidents such as the *Exxon Valdez* oil spill off Alaska in 1989 and the *Prestige* sinking off the coast of north-west Spain in 2002

- on air, where poor quality is responsible for ill health on a local scale and for substantial effects on a global scale, where global warming is seen to be the main result (diagram **D**).

## ■ Causes of climate change and global warming

Natural climate change is thought to be the result of various shifts and cycles associated, for example, with slight changes in the earth's axis or orbit around the sun. Many scientists believe that the recent trend of global warming is to some extent caused by the actions of people.

One important natural function of the atmosphere is to retain some of the heat lost from the earth. This is known as the **greenhouse effect** (diagram **D**). Without this 'blanketing' effect it would be far too cold for life to exist on earth.

Just like a greenhouse, the atmosphere allows most of the heat from the sun (short-wave radiation) to pass straight through it to warm up the earth's surface. However, when the earth gives off heat in the form of long-wave radiation, some gases such as carbon dioxide and methane are able to absorb it. These gases are called **greenhouse gases** (table **E**). In the same way that panes of glass keep heat inside a greenhouse, the greenhouse effect keeps the earth warm.

**C**  *Smog in Beijing, home of the 2008 Olympics, before the games*

### Did you know ??????

At the opening ceremony of the Beijing Olympics on 8 August 2008, a British smog monitoring team had to close down its website after it recorded an air pollution index (API) of between 101 and 150 micrograms per $m^3$ (100 is regarded as safe), whereas the official Chinese figures showed 95.

### ∞ links

The Environment Agency gives extensive advice at **www. environment-agency.gov.uk**.

You can investigate air quality at **www.airquality.co.uk** and global warming at **www. worldviewofglobalwarming.org**.

### Key terms

**Greenhouse effect**: the blanketing effect of the atmosphere in retaining heat given off from the earth's surface.

**Greenhouse gases**: gases such as carbon dioxide and methane, which are effective at absorbing heat given off from the earth.

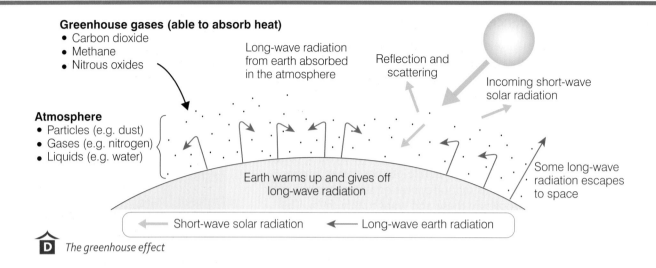

**Greenhouse gases (able to absorb heat)**
- Carbon dioxide
- Methane
- Nitrous oxides

Long-wave radiation from earth absorbed in the atmosphere

Reflection and scattering

Incoming short-wave solar radiation

**Atmosphere**
- Particles (e.g. dust)
- Gases (e.g. nitrogen)
- Liquids (e.g. water)

Earth warms up and gives off long-wave radiation

Some long-wave radiation escapes to space

Short-wave solar radiation ← Long-wave earth radiation

**D** *The greenhouse effect*

**E** *Greenhouse gases*

| Greenhouse gas | Sources |
|---|---|
| **Carbon dioxide:** accounts for an estimated 60 per cent of the 'enhanced' greenhouse effect. Global concentration of carbon dioxide has increased by 30 per cent since 1850 | Burning fossil fuels (e.g. oil, gas, coal) in industry and power stations to produce electricity, car exhausts, deforestation and burning wood |
| **Methane:** very effective in absorbing heat. Accounts for 20 per cent of the 'enhanced' greenhouse effect | Decaying organic matter in landfill sites and compost tips, rice farming, farm livestock, burning biomass for energy |
| **Nitrous oxides:** very small concentrates in the atmosphere are up to 300 times more effective in capturing heat than carbon dioxide | Car exhausts, power stations producing electricity, agricultural fertilisers, sewage treatment |

## ■ Human activities and global warming

In recent years, the levels of greenhouse gases (principally carbon dioxide and methane) have increased (graph **F**) and many scientists believe that this is the result of human activities such as the burning of fossil fuels in power stations, the dumping of waste in landfill sites and burning trees during deforestation.

It is the increased effectiveness of the greenhouse effect that scientists believe is causing global warming. For the first time in history, human activities appear to be affecting the atmosphere with potentially dramatic effects on the world's climate. In 2007 the IPCC stated that global climate change is 'very likely' to have a human cause. It made a prediction that by the end of the 21st century temperatures would rise by between 1.8°C and 4°C and that this would lead to a sea-level rise of 18–59cm.

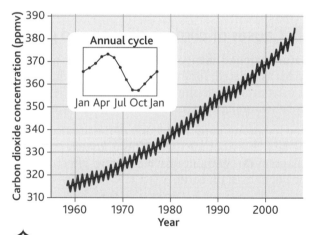

**F** *Increases in carbon dioxide obtained from direct readings at the Mauna Loa Observatory, Hawaii*

## ■ Impacts of global warming in the world

Table **G** lists some of the advantages and disadvantages that global warming may have. It is important to remember that these impacts are highly speculative and based on computer models, which may turn out to be inaccurate. Always consider these impacts to be 'possible' rather than 'probable'. Nonetheless, take note that global warming could have very severe impacts indeed (map **H**).

### ∞ links

More about global warming can be found at **www.defra.gov.uk/ environment/climate**.

**G**

| Advantages of global warming | Disadvantages of global warming | |
|---|---|---|
| ■ Frozen regions of the world such as Siberia and northern Canada may be able to grow crops in a milder climate | ■ Higher sea levels may flood low-lying areas such as Bangladesh, Myanmar and the Netherlands, threatening the lives of 80 million people | ■ Tropical storms affecting the Caribbean and the USA may increase in magnitude |
| ■ Canada's North-west Passage may become ice-free and can be used by shipping | ■ Islands such as the Maldives and Tuvalu may completely disappear as sea levels rise | ■ Loss of glaciers (fresh water) in the Himalayas may threaten agriculture and water supply in India, Nepal and China |
| ■ Energy consumption may go down as temperatures increase in densely populated parts of the world such as north-west Europe | ■ Parts of Africa may become drier and more prone to droughts, leading to starvation and civil war | ■ Hazards such as landslides, floods and avalanches may become more common in mountainous areas such as the Alps |
| ■ Fewer deaths or injuries due to cold weather | ■ Cereal yields are expected to decrease in Africa, the Middle East and India | ■ Arctic ice may melt completely |
| ■ Longer growing season in rich agricultural areas such as Europe and North America will increase food production | ■ An additional 220–400 million people may be at risk from malaria, particularly in China and central Asia | ■ Some species, whose habitat changes, may become extinct – there is considerable concern about polar bears in the Arctic |
| | | ■ Alpine ski resorts may be forced to close due to lack of snow |

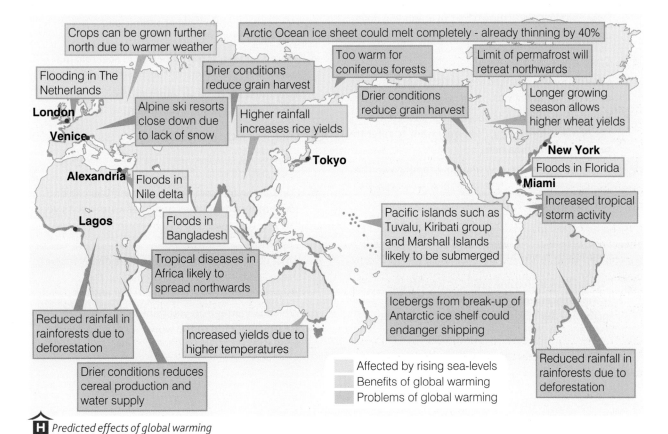

**H** *Predicted effects of global warming*

## Acid Rain

Acid rain is produced by the burning of fossil fuels. It is formed when emissions of sulphur dioxide and nitrogen oxides react in the atmosphere with water and oxygen to form various acidic compounds. These compounds then fall to the ground in either wet or dry form (diagram **I**). Areas affected include most of eastern Europe from Poland northward into Scandinavia, the eastern third of the United States, and southeastern Canada.

Electric power plants account for about 70 per cent of sulphur dioxide emissions and about 30 per cent of nitrogen oxide emissions. Cars, trucks and buses also are major sources of nitrogen oxides.

Acid rain acidifies lakes and streams and contributes to damage of trees. This was noticed first in Scandinavian countries. Hundreds of lakes became too acidic to support sensitive fish species. In addition, acid rain accelerates the decay of paints and buildings.

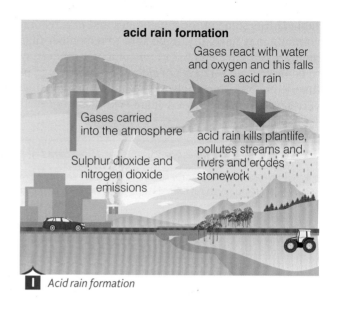

**acid rain formation**

Gases react with water and oxygen and this falls as acid rain

Gases carried into the atmosphere

Sulphur dioxide and nitrogen dioxide emissions

acid rain kills plantlife, pollutes streams and rivers and erodes stonework

**I**  *Acid rain formation*

## Activities

**1**  Study table **A** on page 152.

a  Draw a line graph to show the total world population figures from 1750 to 2050.

b  Describe the trends shown by the graph you have drawn.

**2**  Study graph **B** on page 152.

a  Summarise the changes in population in the different world regions.

b  Explain how the trends account for an increase in the demand for energy. (You may want to refer to section 7.4 on China here.)

**3**  a  List the products you and your family own that use energy.

b  Ask your parents to highlight the items they would not have owned 10 and 20 years ago.

c  How will the amount of energy used by you and your family have changed?

**4**  Explain how increasing levels of wealth and technological advances have resulted in an increasing demand for energy. Try to illustrate your answer using your own experience.

**5**  Study the text above and diagram **I**.

a  Using diagram **I** explain the cause of acid rain

b  Why is eastern Europe badly affected by acid rain?

c  How can acid rain have economic and environmental effects?

d  There has been some reduction in acid rain in recent years. How might this have been achieved?

e  Why must there be an international solution to the problems caused by acid rain?

**6**  Study table **G** and map **H** on page 155.

a  Present the information in table **G** as labels on a blank outline map of the world.

b  Conduct an internet search to see if you can find some more advantages of global warming. As you can see, there are not very many in table **G** and map **H**.

c  What do you consider to be the three disadvantages that could have the greatest impact on the geography of the world? Explain your answer.

d  Which advantage do you think will bring the greatest benefits to the world? Explain why you think this.

# 7.6   How can energy use be sustainable?

## ■ The importance of international directives

Air pollution knows no bounds. It does not stop at international borders and therefore cooperation is needed between countries worldwide if air quality and global warming issues are to be effectively addressed. The Earth Summit in Rio de Janeiro in 1992 marked the first real international attempt to cooperate to reduce emissions. Richer countries agreed there would be no increase in emissions.

The **Kyoto Protocol** in 1997 went further, with an agreement by industrialised countries to reduce greenhouse gas emissions to 5 per cent below 1990 levels between 2008 and 2012. EU countries as a whole should show an 8 per cent reduction, but individual countries have their own targets. The treaty became legally binding in 2005, when enough countries responsible for 55 per cent of the total emissions had signed. The USA has declined to sign the agreement but by August 2011 there were 191 signatories. The poorer nations, including those with many industries, do not have to reduce their emissions. Countries can trade in their carbon credits – the amount of greenhouse gases they are allowed to emit. Countries putting more pollution into the atmosphere than they should can buy carbon credits from a country below its agreed level.

The Bali Conference in December 2007 sought to establish new targets to replace those agreed at Kyoto. No figure was decided, only a recognition that there would need to be 'deep cuts in global emissions'.

The Durban Conference in December 2011 agreed to a legally binding deal comprising all countries, including USA, China and India, which will be prepared by 2015, and take effect in 2020.

### Local initiatives

The phrase 'think globally, act locally' indicates the need for individuals and groups to seek to reduce pollution and to take responsibility for this. Reducing the use of resources not only increases their life, it also reduces pollution and energy in production. We can seek to seize the initiative by conservation and recycling, and therefore reducing waste and the need for landfill.

### Conservation

Conservation can involve simple things like turning off lights and appliances when they are not being used, filling a kettle with only the water that is needed rather than to the top and buying reusable carrier bags rather than accepting free plastic bags.

### Recycling

Local authorities provide a variety of recycling containers for paper, cans, glass, plastic, cardboard and garden waste, and many encourage composting in an attempt to reduce waste thrown into bins. This in turn reduces the amount that is put into landfill. Figures from the Department for Environment, Food and Rural Affairs (Defra) show that in 2009–10, 40.1 per cent of household waste was recycled. By 2015 the government want only 35 per cent of waste to go into landfill sites.

**Key terms**

**Kyoto Protocol**: an international agreement to cut $CO^2$ emissions to help reduce global warming.

**Biofuels**: the use of living things such as crops like maize to make ethanol (an alcohol-based fuel) or biogas from animal waste. It is the use of crops that has become especially important.

**A** *Burbo Bank wind farm, Liverpool Bay*

## The use of renewable energy

Renewable energy promises sustainable development. Our reliance on fossil fuels to meet our energy needs is only short term. Supplies of coal, oil and natural gas are limited and renewable alternatives have been used for some time, albeit on a relatively small scale. Hydroelectric power (HEP), solar, tidal and wind power are sustainable options (fact file **C**). They share similar advantages over the use of fossil fuels, but are not without their opponents. The recent surge in the use of **biofuels** also appears at first to offer this possibility, but there is debate about this.

Fossil fuels make up a large proportion of the world's energy mix (figure **B**). They are non-renewable and finite i.e. they will eventually run out. At current rates of use, it is likely that oil supplies will run out within 60 years and gas supplies within 80 years.

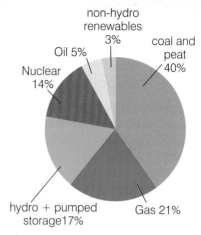

**B** A global energy mix 2010

| | Fact file |
|---|---|

### Wind power in the UK

- **Aims**: to generate 10 per cent of power by renewable energy sources.

- **Role**: to be responsible for one-third of electricity generated.

- **Number of wind farms and turbines operational in July 2011**: 306.

- **Location**: mostly onshore.

- **The future**: offshore wind farms to become more important, with 8 GW to be generated by 2016.

- **Size of a modern wind turbine**: 100 to 120 m, including the blades. Offshore wind turbines tend to be the largest. For comparison, the London Eye is 135 m high.

- **Offshore operations**: these are usually within 4 to 6 km off the coast, but some planned developments will be over 18 km.

- **Location requirements**: the shallow waters off the coast of the UK are an advantage. Wind farms require an exposed location, whether onshore or offshore, clear of any obstructions (such as buildings onshore). Small differences in distance can mean a real difference in the potential of the site, e.g. a site that is 10 per cent less windy means 20 per cent less energy is generated.

**D** The location of wind farms in the UK

- **Opinions – against**: noise levels for a distance of 350 m are measured at 35 to 45 dB, while a busy office is measured at 60 dB. Research suggests that turbines are avoided by migrating birds. House prices nearby may reduce. The cost of generating electricity is more expensive than traditional methods (more than double per kW hour). Turbines can be very tall and an eyesore.

**Opinions – for**: wind is free and can now be captured efficiently. The energy it produces does not cause green house gases or other pollutants. Although wind turbines can be very tall each takes up only a small plot of land (a small footprint). This means that the land below can still be used, for example for agriculture. At sea the turbines can attract fish and have no long term affects on sea life. Some people find wind farms an interesting feature of the landscape. Wind turbines can be used to produce electricity in remote areas that are not connected to the National Grid. Individuals could also choose to generate their own energy.

## BASF – a carbon negative company

BASF has set goals to cut its greenhouse gas emissions by 25 per cent by 2020. Progress is already being made (Diagram **E**).

As well as cutting down on its own emissions, BASF provides environmentally friendly products to other companies. It says these products can save three times the air pollution given off while they are being made. This is achieved by making:

- thermal insulation products to reduce heat loss from buildings
- lighter plastics to reduce the weight of cars and reduce fuel consumption.

BASF has also developed:

- technology to cut emissions of greenhouse gases
- technology to cut down on releases of gases from agriculture
- alternative energy projects
- biodegradable plastics.

BASF is working with less developed countries to cut down on environmental damage. In Ethiopia it has planted 15,500 hectares of forest to reduce soil erosion and improve water quality. In Honduras it has built a hydroelectric power station to provide people with energy, cut down on fossil fuel use and create jobs. BASF is also contributing to the Millennium Development Goals, for instance by starting education programmes in Brazil, benefiting 10,000 children.

66 *Climate change is one of the key challenges facing society. BASF has taken up the challenge and offers a variety of solutions to help protect our climate.* 99

*Eggert Voscherau, Vice-Chairman of BASF SE*

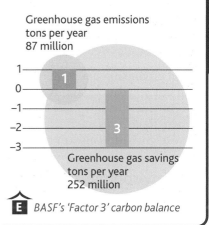

Greenhouse gas emissions tons per year 87 million

Greenhouse gas savings tons per year 252 million

**E** *BASF's 'Factor 3' carbon balance*

## ∞ links

You can investigate biofuels by searching **www.bbc.co.uk**. This website is also useful for information on offshore wind farms.

The British Wind and Energy Association website at **www.bwea.com** gives lots of information on wind farms in the UK.

Local authority websites provide information on recycling and waste disposal.

### Activities

**1** Study map **D**.
a Describe the distribution of wind farms shown on the map.
b Explain reasons for the distribution you have described.

**2** Different people have contrasting views on the development of wind farms such as Burbo Bank (photo **A**) and onshore wind farms.
a Choose one of the following who you think would support wind farms and one who you think would oppose them.
  i a resident overlooking the new development in favour of renewable energy
  ii a NIMBY ('Not In My Back Yard') supporter of wind energy
  iii an electricity-generating company director
  iv a member of the Royal Society for the Protection of Birds (RSPB)
  v a member of Greenpeace for action against climate change
  vi a resident who wants cheap electricity.
b From the point of view of each person, give your views explaining fully why you are for or against the development of wind farms.

**3** a Explain the importance of an international approach to reduce air pollution.
b In your view, how successful have international initiatives been? Provide evidence to back up your answer.

**4** Use your own experience of where you live and research what your local authority does to encourage energy conservation. Work in pairs. You have been commissioned by your local council to produce a leaflet encouraging people to conserve energy. Design and present your leaflet so that it is informative, clear, well illustrated and eye-catching.

# 7.7 What are the effects of increasing global food production?

Food production has increased significantly throughout the world in the last 50 years (graph **A**). The quest to increase food production worldwide – whether out of necessity for survival or to sell to areas where food items are needed – has had significant impacts. These can be categorised as environmental and socio-economic.

## ■ Environmental impacts

Transporting food longer distances – the idea of **food miles** – increases our **carbon footprint**. People in the UK demand out-of-season produce and this demand is met by importing food. One half of vegetables and 95 per cent of fruit comes from abroad out of season. We import 1 per cent of food by air, but this accounts for 11 per cent of carbon emissions resulting from transportation of UK food. Therefore, the more we rely on imported food – especially air-freighted food – the more we are contributing to air pollution and global warming.

Whilst it may seen logical to suggest that food flown thousands of miles to the UK is bad for the environment, it's not quite so simple. For example, Kenyan green beans are produced efficiently using manual labour and natural fertilisers. British beans are grown in oil heated greenhouses or grown in fields sprayed and ploughed by tractors. According to Gareth Thomas, Minister for Trade and Development (2008–2010): 'Driving six and a half miles to buy your shopping emits more carbon than flying a pack of Kenyan green beans to the UK.'

Apples are harvested in September and October. Some are sold fresh while the rest are chill stored. For most of the following year, they still represent good value (in terms of carbon emissions) for British shoppers. However, by August those Cox's orange pippins and Braeburns will have been in store for 10 months. The amount of energy used to keep them fresh for that length of time will then overtake the carbon cost of shipping them from New Zealand. It is therefore better for the environment if UK shoppers buy apples from New Zealand in July and August rather than those of British origin.

As populations increase in certain areas of the world, there is pressure to increase food production. This may involve farming on marginal land that is not really suitable. Photo **B** illustrates the difficult conditions in which people try to grow food in order to survive. Here, the already poor-quality land is likely to become even poorer. As the meagre crops are harvested, no goodness is returned to the soil and so it becomes exhausted. The lack of vegetation cover makes the area prone to soil erosion, where it is easily washed or blown away (photo **C**). This is known as **environmental degradation**.

**In this section you will learn**

how efforts to meet the global demand for food can have both positive and negative effects

how these impacts can be classified as environmental socio-economic.

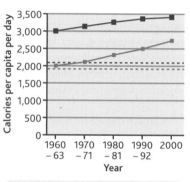

 **A** *Global food production*

**Key terms**

**Food miles**: the distance that food items travel from where they are grown to where they are eaten.

**Carbon footprint**: the amount of carbon generated by things people do, including creating a demand for out-of-season food.

**Environmental degradation**: undesirable changes to the natural environment through the removal of natural resources and disruption to natural ecosystems. Human activity is a major cause.

**B** *Marginal land in Darfur*

**C** *Soil erosion caused by growing crops on marginal land*

## Socio-economic impacts

Social issues relate to aspects such as health, safety and quality of life, and include the right to a clean water supply. Growing **cash crops** for a source of income can offer economic benefits. However, socially there are potentially more problems. The areas around Lake Naivasha in Kenya and north of Mt Kenya are home to the profitable flower industry (diagram **D**).

> **Key term**
>
> **Cash crops**: crops grown in order to sell to make a financial profit.

### ∞links

See spread 10.7 on p.240 for more on flower production in Kenya

| | |
|---|---|
| Water supplies are affected by the fertiliser | Some of the fertiliser is washed into soils and seeps underground |

River Ngiro in the north has sections without water

Local farmers say the flower growers are taking more water than they should legally and leaving them short

Population around Lake Naivasha increased from 50,000 to 250,000 as people sought work in the greenhouses or in the fields

Fertile land is used for growing flowers and not for food

An experienced worker can cut 1,000 roses an hour

Chemicals sprayed frequently on the flowers cause rashes and chest problems

Water levels are falling – a settlement of 20,000 people could be supplied with water that is used by the flowers

The Maasai are especially struggling for water to the east and north of Mt Kenya

**D** *Growing flowers in Kenya*

Growing cash crops as well as food crops is often the way forward for many small-scale farmers – some even seek to sell any surplus of food they produce. Additional cash allows investment in the farm and other items to be bought. However, there are problems. There is often a need to intensify production, which means increasing the use of fertilisers and pesticides. These cost money and a vicious circle can be set up creating **rural debt** (diagram **E**).

## Buying locally produced food

In the UK, farmers benefit from local people buying food they produce. Buying from local, or indeed regional or national sources, should benefit the domestic farming industry and address environmental concerns. Others might argue from a different perspective. What would the effect on Kenya be if the UK stopped importing strawberries, for example?

In the UK, we can ensure that we support local produce by:

- looking at labels in supermarkets, which increasingly give the specific origin of foods
- visiting specialist local shops
- buying online from 'local' producers
- supporting local farmers' markets (photo **F**)
- attending regional agricultural shows, which celebrate and sell local produce
- heeding the advice of celebrity chefs such as Gordon Ramsay who joined the debate in May 2008 by saying 'There should be stringent licensing laws to make sure produce is only used in season. I don't want to see strawberries from Kenya in the middle of March. When we haven't got it, we take it off the menu.'

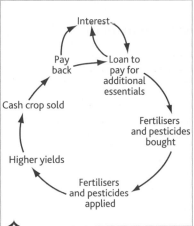

**E**   *The effects of borrowing money*

## ∞ links

The case study on 'Flower growing in Kenya' on **www.learningafrica.org.uk** is particularly useful, as is **www.farmersmarkets.net**.

### Activities

1   Working in small groups, choose at least two different types of non-seasonal fruit and vegetables each. Visit one (or more) supermarkets, making sure in advance that they stock the produce you have decided on. If not, you will need to change your choice of items.

a   Write down the origin of the items you have selected and whether they have been brought in by air. Many supermarkets now give this information.

b   Display the origins of the fruit and vegetables on a world map.

c   Summarise what you have discovered about the origin of fruit and vegetables eaten in the UK.

d   Describe the environmental impact of this.

2   Study photos **B** and **C** on page 161.

a   Give the meaning of 'marginal land'.

b   Use the photos to illustrate the difficulties of farming on marginal land.

c   Use photo **C** to explain the negative effects of farming on marginal land.

**F**   *Fresh, local produce from a farmers' market*

# 8 Contemporary issues in tourism

## 8.1 Why has global tourism grown?

### Growth in tourism

Tourism is the world's largest industry, worth $500 billion dollars in 2007. There were 940 million international tourist arrivals in 2009 and this is set to rise to a massive 1.6 billion by 2020. In most countries, domestic tourism (people going on holiday in their own country) is between four and five times greater than international tourism.

**A** *International tourist arrivals (millions)*

| Region | 2006 | 2010 | Percentage change 2006–10 | Percentage of world tourism (2010) |
|---|---|---|---|---|
| Africa | 50 | 49 | −2.0% | 5.2% |
| Americas | 136 | 151 | +11.0% | 16.1% |
| Asia and Pacific | 168 | 204 | +21.4% | 21.8% |
| Europe | 461 | 471 | +2.2% | 50.5% |
| Middle East | 41 | 60 | +46.3% | 6.4% |
| World | 856 | 935 | +9.2% | 100.0% |

**In this section you will learn**

the many types of landscapes and holidays that attract people and why

reasons for the global growth of tourism.

**B** *Tourism in the Caribbean*

**Study tip**

You should be able to interpret tables of statistics. Usually, we look for patterns or trends and drawing a graph can be the easiest way to spot these. Sometimes there are few differences in the data. Nevertheless, tiny differences can still be important and may show a trend, and this is what you need to pick out.

The tourism industry is one of the greatest providers of jobs and income in all countries. Many countries rely heavily on income from tourism. Caribbean countries get half their GDP from tourism. The top six tourist destination countries are France, Spain, the USA, China, Italy and the UK. Germans spend more per person than any other nation on holiday, followed by Americans, British, French and Japanese.

### Factors affecting tourism's growth

Growth in tourism is explained by three sets of factors.

#### 1 Social and economic factors

Since the 1950s people have become wealthier. Incomes are higher and more money is available for luxuries such as tourism. Most families have two working parents whereas in the past it was usually one.

People have more leisure time. Holiday leave time has increased from two weeks per year in the 1950s to between four and six today. Life expectancy has risen so more people are retired. Many have good pensions and can afford several trips a year. They also have more time to travel.

## 2 Improvements in technology

Travel today is quick and easy – motorways, airport expansion and faster jet aircraft have all contributed to this. Flying has become cheaper and booking online is quick and easy. In 2008 the rapid rise in oil prices had an impact on the cost of flights and more people took domestic holidays to save money.

## 3 Expansion of holiday choice

During the 1950s and 1960s coastal resorts were popular and in the UK the National Parks were opening and offering new opportunities. The 1970s saw a decline in seaside holidays due to competition from cheap package holidays to mainland Europe, especially Spain. Packages are now available to destinations all over the world that offer a huge variety of sights and activities. Ecotourism and unusual destinations such as Alaska are expanding rapidly.

**C** *Faster and larger aircraft have made international travel quicker and cheaper*

## ◼ Tourist attractions

Many people choose to visit cities to enjoy the culture associated with museums, art galleries, architecture or shops and restaurants. Cities such as London, Rome and Paris have a huge amount to offer tourists of every age. The natural landscape is also a major 'pull' for tourists, particularly mountains such as the Alps in Europe or the beautiful stretches of coastline found in the Mediterranean or the Caribbean.

## Activities

**1** Study table **A** on page 163.

a Which continent had the greatest tourism business in 2010?

b Why do you think this continent has such a large tourist industry?

c Which continent had the smallest tourism business in 2010?

d Why do you think this continent has not yet developed its tourism as much as other regions?

e Which continent's tourism grew by the largest percentage between 2006 and 2010?

f Which continent's tourism grew the least?

g Is there a relationship between scale of tourism and rate of growth? Explain your answer.

**2** Study photos **D**, **E** and **F** opposite.

a Draw a field sketch of one of the photos. Label all the pull factors that might attract people to come to that region of Italy.

b What types of people are most likely to be attracted to each area? Remember to consider the seasons. Some landscapes are more attractive in winter than in summer.

c Italy is a base for holidays all year round. Explain why this is so.

d For the three Italian holiday types shown, assess their tourist potential for members of your family and/or friends. Give your reasons. Who would be most attracted to what and why?

## Italy

Italy is a country with a great variety of landscapes. Photos **D**, **E** and **F** show the three key types: mountains, cities and coastline. All the places shown have a busy tourist business, making an important contribution to the national economy. Venice is well known for its canals and Renaissance architecture; Florence for its art galleries. Skiing in the Alps and sunbathing on the coast are popular with both Italians and visitors from other countries.

The example of Italy illustrates point 3 above – expansion of holiday choice. People want to visit a greater variety of places and the tourist industry grows and adapts to supply what the market demands.

**D** *The Italian Alps*

**E** *Venice is a popular tourist destination*

**F** *Coastal tourism in Vernazza*

### Study tip

Photos can be labelled directly with arrows to point out the features. There needs to be space around each photo because the labels need to be detailed. Keep what you write clear, straightforward and to the point.

# 8.2 How important is tourism in different countries?

Tourism is an important part of the economies of many richer countries, especially those in Western Europe and North America. Today, it is increasingly seen by developing countries as one of the best ways to earn foreign income, provide jobs and improve standards of living. Countries want to take advantage of the growing numbers of tourists and the money they have to spend.

> **In this section you will learn**
>
> the economic importance of tourism to countries in contrasting parts of the world
>
> how to compare and contrast tourist regions.

## The economic importance of tourism

Tables **A** and **B** show the top 10 countries for tourist receipts (national income from tourism) and most tourist arrivals (number of tourists every year).

- France has had more tourists than any other country for many years. French tourism includes every type of holiday such as city breaks, holiday cottages, camping and skiing.

- The USA earns more than any other country from tourism, and has the second largest number of visitors. Europeans consider a trip to the USA as more special than staying in Europe, so they are likely to stay longer and spend more.

- China is high in both tables **A** and **B**. For many people, distance makes it too expensive a place to visit, but its variety of unusual landscapes and unique culture attracts increasing numbers with both time and money. This trend is likely to continue.

- In the Caribbean almost 50 per cent of visitors come from the nearby USA, with France, Canada and the UK also important sources of business. Expenditure per tourist ranges between $324 per holiday in Belize to $2,117 in the Virgin Islands, which attracts the wealthiest visitors.

Essential jobs are created in all countries, but the contribution that tourism makes to GDP varies greatly between wealthier and poorer countries. Rich countries have a broadly balanced economy, of which tourism is one part. On the other hand, in less well-off countries tourism can be essential. In the Caribbean, for example, several small island countries rely heavily on tourism to provide national income and employment. Around 80 per cent of Barbados's national income comes from tourism.

**A** *Countries with the largest tourist receipts, 2009*

| Country | Annual tourist income ($ billions) |
|---------|-----------------------------------|
| USA | 110.1 |
| Spain | 61.6 |
| France | 55.6 |
| Italy | 45.7 |
| China | 40.8 |
| Germany | 40.0 |
| UK | 36.0 |
| Australia | 24.7 |
| Turkey | 22.0 |
| Austria | 21.8 |

**B** *Countries with the most tourist arrivals, 2009*

| Country | Number of tourists (millions) |
|---------|-------------------------------|
| France | 78.5 |
| USA | 57.9 |
| Spain | 57.3 |
| China | 53.1 |
| Italy | 42.7 |
| UK | 30.1 |
| Ukraine | 25.5 |
| Turkey | 25.0 |
| Germany | 24.9 |
| Russia | 23.7 |

## Benefits of tourism in poorer countries

- Many people are employed to serve tourists such as waiters (photo **C**), souvenir shop assistants and tour guides. In Antigua and Barbuda 30 per cent of the population work in these jobs, but in Jamaica only 8 per cent.

- Tourists spend their holiday money in pounds sterling, US dollars or euros. This foreign exchange is essential to poorer countries. It can be used to buy goods and services from abroad.

- Many governments tax visitors to help pay for the extra services they use such as water supply, drainage, electricity and roads.

- Extra jobs are created indirectly. Hotels buy some produce from local suppliers to feed the visitors.

- Many small businesses have been started up to serve the tourists themselves and supply the services they demand. These include taxis, bars and restaurants, builders and maintenance workers.

**C**  *Working as a waiter*

### Activities

1  Study tables **A** and **B** on page 166.

a  Draw a pie chart using the data in table **A**. Calculate the percentage that each country/region is out of the total figure. 100 per cent represents 360 degrees, so multiply each percentage you have calculated by 3.6 to obtain the number of degrees for each sector in your pie chart. Label each sector with the appropriate country/region or add a key to explain your shading. The data in the table is in size order. If you follow this in your chart, starting at the top, it will be easier to read and interpret.

b  Which are the top three countries for tourist receipts?

c  Which are the top three countries for tourist arrivals?

d  Comment on any similarities and differences in your answers to b and c.

e  Compare the balance of richer and poorer countries in the two tables.

# 8.3 How do we manage tourism in the UK?

## The growth of tourism in the UK

Almost all UK tourism used to be domestic – British people holidaying in the UK. Only the wealthy and privileged were able to go abroad.

Domestic tourism grew quickly in the 1950s and 1960s as the growing UK economy provided higher pay and more time off work. Having an annual holiday became common. UK seaside holidays peaked in the early to mid-1970s, with 40 million visitors annually (graph **A**). After that, Britain's seaside resorts declined as package holidays abroad grew in number and affordability.

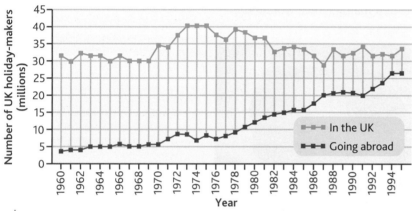

**A** Numbers of UK residents holidaying in the UK and abroad, 1960–95

**B** British seaside resorts were popular in the 1960s

Between the world wars, wealthier people began to go abroad. Cheap package holidays with guaranteed hot weather attracted people of all incomes. It was often cheaper to go to a Spanish resort like Benidorm than to holiday in the UK. British weather was seen as too unreliable. Many small coastal hotels were forced to survive by housing the homeless during the 1980s, decreasing their reputation even more. In 2010–11 UK residents made 55.7 million trips abroad.

**Did you know** ❓❓❓❓❓

The banking crisis of 2008–09 meant people had less money to spend. People reduced the number of holidays they took while they waited for the economic situation to improve.

## The contribution of tourism to the UK economy

The UK economy earns over £114 billion (2008) every year from tourism and leisure. This amount usually grows slightly annually. Around 27.7 million overseas visitors spend over £13 billion of this sum. Restaurants and hotels make up a large proportion of these earnings, at £20 billion and £16 billion respectively. More than 100 new hotels opened in the UK between September 2004 and December 2005, creating more jobs and income. The London Eye is the most visited paying attraction in the UK, with 3.7 million visitors each year. An estimated 1.1 million tourists travelled to London for the Royal Wedding in 2011.

**C** The Royal Wedding, 2011

# The Butler tourist resort life-cycle model

This **life-cycle model** says that any tourist resort starts on a small scale, develops into something more significant, then either goes into decline or makes changes to maintain its attractions (graph **E**). There are six stages.

## 1 Exploration

Small numbers of visitors are attracted by something particular, e.g. good beaches, attractive landscape, historical or cultural features. Local people have not yet developed many tourist services.

## 2 Involvement

The local population sees the opportunities and starts to provide accommodation, food, transport, guides and other services for the visitors.

## 3 Development

Large companies build hotels and leisure complexes and advertise package holidays. Numbers of tourists rise dramatically. Job opportunities for local people grow rapidly, but this brings both advantages and disadvantages.

## 4 Consolidation

Tourism is now a major part of the local economy, but perhaps at the expense of other types of development. Numbers of visitors are steady making employment more secure. However, some hotels and other facilities are becoming older and unattractive, so the type of customers attracted goes downmarket. Rowdiness becomes a problem.

## 5 Stagnation

The resort becomes unfashionable and numbers of visitors start to decline. Businesses change hands and often fail.

## 6 Decline or rejuvenation

Decline: visitors prefer other resorts. Day trippers and weekenders become the main source of income.

Rejuvenation: attempts are made to modernise the resort and attract different people to enjoy new activities.

Blackpool is a good example of a resort reinventing itself. Day trippers and weekenders now bring in most of the income, although websites and brochures make a huge effort to attract people for longer periods.

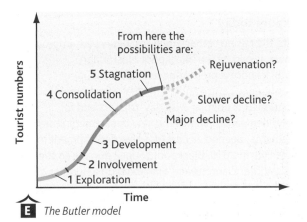

**D**  *The Mediterranean coastal resort of Kalami on the Greek island of Corfu. In what stage of the Butler model would you place this resort and why?*

**E**  *The Butler model*

Tourist numbers / Time

5 Stagnation
4 Consolidation
3 Development
2 Involvement
1 Exploration

From here the possibilities are:
Rejuvenation?
Slower decline?
Major decline?

**Key term**

**Life-cycle model**: a theoretical model used to describe the changes that take place as a tourist resort develops.

**Did you know** ??????

In 2009 there were 30.1 million visits to the UK. Visitors spent over £15 billion.

**Study tip**

The Butler tourist resort life-cycle model has six stages (with stage six having two options). Learn these by means of a mnemonic: **E**dinburgh **I**s **D**eveloping **C**astle **S**ites **D**aily (or **R**egularly).

# Blackpool, a UK coastal tourist resort

## Blackpool's growth and stagnation

Located on the Lancashire coast in the north-west of England, Blackpool became a major tourist centre during the 19th century to serve the inhabitants of northern industrial towns. Factory workers could increasingly afford a holiday, travelling by train to the nearby coast.

Blackpool boomed between 1900 and 1950. However, as people's disposable income increased, they preferred to try somewhere different. Soon package holidays abroad created huge competition for Britain's seaside resorts. Blackpool's summer weather can be unreliable, which proved a disadvantage.

**F**   *Factory workers on holiday, 1951*

## Responding to the decline

As one of the larger resorts, Blackpool did attract some private investment and local authority grants to upgrade hotels, turn outdoor pools into indoor leisure centres and increase car-parking provision. Many smaller failing hotels were converted into self-catering holiday flats. Blackpool's attractions still made it a little different to other resorts. The famous Blackpool Tower, modelled on the Eiffel Tower in Paris, gives fantastic views up and down the Lancashire coast and inland towards the Pennines. The complex there includes the Tower Ballroom, famous for national ballroom dancing competitions, and the Tower Circus. The town upgraded its zoo and a Sealife Centre was built. The

Blackpool Illuminations – a light show stretching along the Golden Mile (the central section of the sea front) – began in 1879 and has been upgraded several times with advancing technology.

**G**   *Blackpool Tower and the Illuminations*

Blackpool should have been quicker to fight the competition from package holidays. Eventually it lost much of its family holiday business and came to rely on day trippers and stag and hen party business – not popular with residents and bad for the town's image.

## The supercasino

One approach taken by Blackpool to get out of recession was to apply to the government to be the home of the UK's first supercasino, a huge leisure and entertainment complex based on those in Las Vegas and Atlantic City in the USA. Blackpool was the favourite applicant on the shortlist. The proposed sea-front site included a conference centre and a wide variety of entertainments. Some 20,000 jobs and £2 billion of investment would have been generated. The town's high 3.6 per cent unemployment rate would have been greatly improved. Nevertheless, not everyone was in favour, but the final decision closed the debate – Blackpool lost the supercasino development to Manchester, coming only third in the vote. The town council was shocked.

## Blackpool today

Despite the casino setback millions were spent improving the town as part of the 'Blackpool Masterplan'. Today Blackpool is promoting itself as a shopping and conference centre and as an ideal 'short break' destination. It is popular with day-trippers and for stag and hen parties. In 2008 a new department store (Debenhams) opened as part of a new development in Hounds Hill. In 2010, Blackpool council purchased the famous Winter Gardens and embarked on a number of refurbishment projects.

# External factors affecting UK tourism in the early 21st century

Tourism can be limited by political and economic situations. Two key issues have caused difficulties in the early 21st century.

## Terrorism

The destruction of the World Trade Center in New York on 11 September 2001 had a huge impact on travel. The USA stepped up its security overnight, as did the UK and the EU. Airport security checks have multiplied and check-in times increased. London is a terrorist target: the Underground bombing of 7 July 2005 is an example. In the aftermath of such events, visitor numbers decline sharply.

## Exchange rates and the banking crisis

Currency exchange rates control value for money for tourists on holiday. In 2011 the euro was high against the pound, valued at around 87p (compared with 68p previously), so holidaying in France and other Eurozone countries became more expensive. At the same time, £1 was worth $1.60, making the USA a much more attractive holiday destination. Use the internet to compare these currencies today to see how things have changed since 2011.

Since 2008, the world has faced economic difficulties. In the UK many people have faced a pay freeze and the rate of unemployment has increased. The factors have led to a reduction in tourism.

> **Study tip**
>
> Note which aspects of the photos you are being asked to compare and contrast and concentrate only on those. Note what each photo shows in each category. Write down all the similarities and then all the differences.

## Activities

1 Study graph **A** on page 168.
  a In 1960 how many people holidayed within the UK and how many went abroad?
  b Which year had the largest gap between the sets of statistics?
  c Which year had the smallest gap?
  d Describe the trend of holidays taken in the UK.

2 Study graph **E** on page 169. To what extent has Blackpool's development as a tourist resort followed the Butler model? Write at least two paragraphs.

3 Work with a friend to put together some suggestions for making Blackpool attractive to tourists in the future. Use the internet to give you some ideas. Put together a 'Tourism Plan for Blackpool' using ICT if you wish. Don't forget to justify your plan.

4 Use the internet to find photos that illustrate each stage of the Butler Model. Add labels and annotations to support your choice of photo for each stage.

## links

Find out more about Blackpool tourism at **www.visitblackpool.com**.

# 8.4　What is the importance of National Parks for UK tourism?

National Parks are large areas of mainly rural land. The UK's first was the Peak District of Derbyshire, created in 1951 by an Act of Parliament. National Parks aim to conserve natural and cultural landscapes while allowing access for visitors to enjoy them. The first Scottish Park was designated in 2002. In England, the most recently designated National Parks are the New Forest (2004) and the South Downs (2011).

Many National Parks are uplands such as Snowdonia (photo **B** on page 173) and the Lake District. Several National Parks include stretches of coastline (e.g. Pembrokeshire Coast) and a few are in lowland areas (e.g. The Broads). Land remains privately owned (81 per cent), mostly by farmers, but the Forestry Commission, the National Trust, the Ministry of Defence and the water authorities also own some areas. The National Park Authorities only directly control 1 per cent. Local people make their living from the land and tourism provides much-needed jobs.

### In this section you will learn

- the importance of National Parks for UK tourism
- the conflicts caused by tourism in the Lake District
- how to devise management strategies for tourism in the Lake District.

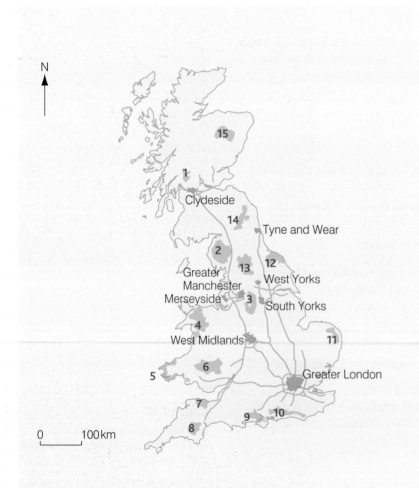

1　Loch Lomond and the Trossachs
2　Lake District
3　Peak District
4　Snowdonia
5　Pembrokeshire Coast
6　Brecon Beacons
7　Exmoor
8　Dartmoor
9　New Forest
10　South Downs
11　The Broads
12　North York Moors
13　Yorkshire Dales
14　Northumberland
15　Cairngorms

- National Park
- Motorway
- Major urban areas

### Key terms

**National Park:** an area where development is limited and planning controlled. The landscape is regarded as unusual and valuable and therefore worth looking after.

**Honeypot site:** somewhere that attracts a large number of tourists who, due to their numbers, place pressure on the environment and people.

**A** The location of National Parks in Britain

## Activities

1 Study map **A**. One of the aims of National Parks is to serve the population with recreation facilities.

a Describe the locations of the National Parks in relation to the major urban areas.

b Based on your answer to **a**, how accessible do you think the National Parks are for most of Britain's population? Make sure you name Parks and cities in your answer.

c Which National Parks are the most accessible? Explain why.

2 Study photo **B**. What types of activities might people undertake in this landscape?

**B** Snowdonia National Park

# The Lake District

## Attractions and opportunities for tourism

The English Lake District is a glaciated upland area in Cumbria, north-west England. It stretches 64 km from north to south and 53 km east to west. It became a National Park in 1951. Famous for its stunning scenery, abundant wildlife and cultural heritage, it is considered to be England's finest landscape.

### The great outdoors

The ribbon lakes and tarns are part of a unique and hugely varied landscape, as well as being a major recreational resource. Lake Windermere specialises in ferry cruises. Most people sail between the main centres of Windermere town (photo **C**) and Ambleside. Small boats are allowed on many lakes. Areas are set aside for windsailing and power-boating so the activities do not clash and quiet areas are left for people seeking peace and quiet. Fishing from the shore or boats is increasing in popularity.

Walking is one of the most popular reasons why people visit the Lake District, whether for a day or longer. Routes vary from short and relatively flat to extremely long and tough. Known as the 'birthplace of mountaineering', even the most experienced climbers find plenty of challenges in scaling back walls of corries and sides of U-shaped valleys (photo **D**). Public access to the fells (open uplands) is unrestricted.

### History and culture

Historical and cultural sites also attract tourists. The Lake District has been occupied since the end of the ice age 10,000 years ago and evidence of early settlement remains in the landscape. The land has been farmed for centuries, leaving a distinctive field pattern with drystone walls. Many 19th-century writers and artists, such as John Ruskin, loved the area. The children's author Beatrix Potter lived at Hill Top on the shores of Lake Windermere (map **E** and photo **G**).

**C** Windermere town centre

**D** Striding Edge, Helvellyn

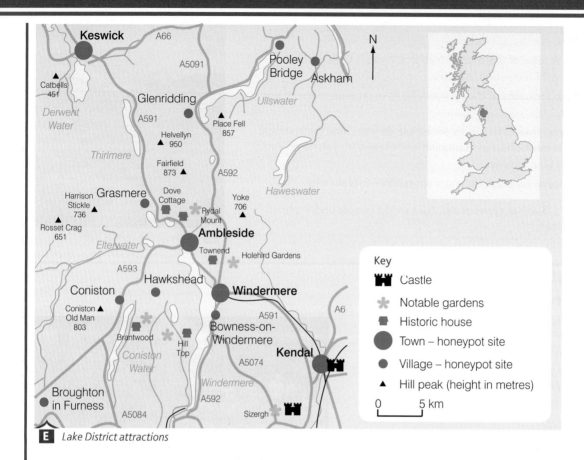

**E** *Lake District attractions*

## Impacts of tourism

There are many times the number of visitors in the Lake District National Park as local population: 12 million tourists (22 million visitor days) to 42,239 residents. The impacts are potentially huge and must be managed as well as possible. There are both negative and positive impacts.

### 1 Traffic problems

Over 89 per cent of visitors come by car, often just for the day. Many roads, including A roads, are narrow and winding. Buses and large delivery vehicles have to use these to service both locals and tourists. Queues are a common problem, especially towards the end of the day when day trippers are heading home.

### Bowness-on-Windermere

Towns like Bowness-on-Windermere were not built originally for the huge volumes of traffic that arrive daily in the summer, especially at weekends. Congestion and parking are serious problems. Bowness has built a new car park at Braithwaite Fold on the edge of town and has extended another, but capacity is still inadequate. In desperation, in the countryside people park on grass verges, causing serious damage.

### 2 Honeypot sites

The Lake District has both physical and cultural **honeypot sites**. Beauty spots, small shopping centres and historic houses all attract hundreds of visitors daily. Catbells (map **E**) is quite an easy climb, so many people walk up this smaller mountain. It therefore suffers from serious footpath erosion (photo **F**). Honeypot sites need to provide access and facilities while remaining as unspoilt as possible.

**F** *Footpath erosion in the Lake District*

**G**  *Hill Top, Beatrix Potter's house – a honeypot site in the summer*

## 3 Pressure on property

Almost 20 per cent of property in the Lake District National Park is either second homes or holiday let accommodation (15 per cent of all housing in 2007). Some local people make a good income from owning and letting such property. The main issues include the following:

- Holiday cottages and flats are not occupied all year.
- Holidaymakers do not always support local businesses, often doing a supermarket shop at home before their trip. On the other hand, the main supermarket in Windermere is often full of visitors buying a great deal of food and drink for their stay.
- Demand for property from outsiders increases property prices in the Lake District. This causes problems for local people who are forced out to find affordable homes on the edge of the region in Kendal or Penrith. This is the most serious tourist problem affecting local communities.

## 4 Environmental issues

Water sports are not allowed on some of the lakes. However, Windermere, the largest lake, has ferries and allows power-boating, windsurfing and other faster and more damaging activities (photo **H**). The main issue is the wash from faster vehicles eroding the shore. Fuel spills are not uncommon, causing pollution.

## Tourism management strategies

In 2007 the Lake District National Park launched its 'Visions for the National Park in 2030'. It aims to maximise economic opportunities in a sustainable way. In 2008 the Cumbria Tourist Board published its new tourism strategy 'Making the Dream a Reality 2008–2018'.

### 1 Traffic solutions

Planning an efficient road network:

- County strategic roads, often dual carriageways, have been built on the edges of the Lake District to help move traffic in and out as efficiently as possible.
- Distributor roads link the small towns and key tourist villages such as Ambleside.
- Develop transport hubs to allow efficient interchange between parking, buses, boats, cycling and walking (for example, Ambleside).
- Traffic on smaller roads has been slowed down by traffic-calming measures in villages, cattle grids in the countryside and an overall maximum speed limit.
- Heavy lorries should be kept off scenic roads.

Planning public transport:

- Where possible bus lanes operate in towns, although narrow streets limit this.
- A park-and-ride scheme called the 'Honister Rambler' operates from Keswick on the edge of the National park. Ramblers are transported to popular sites where parking is a problem, for example, Catbells.
- The use of low-carbon vehicles, for example, buses, have been encouraged.

**H**  *Overcrowding on Lake Windermere*

## 2 Honeypot management

Footpaths:

- The 'Upland Path Landscape Restoration Project (2002-2011)' resulted in the repair of 145 paths which involved creating steps, surfacing with local stone and re-planting native plants.

- A severely eroded path at Whiteless Pike, Buttermere has been repaired using a technique called 'stone pitching'. Large stones are dug into the path to create solid, hard-wearing footpaths. It costs £1,000 per metre.

Parking:

- Roadsides have been fenced off so that cars cannot damage verges.

- Car-park surfaces have been reinforced to prevent damage. 'Waffles' are large concrete slabs with holes in them, like an edible waffle. Soil fills the holes and grass grows, giving a hard green surface.

Litter:

- Bins are provided at popular sites and emptied regularly.

- Signs encourage people to be responsible and reduce litter.

## 3 Property prices

This is the most difficult issue. Management strategies cannot control house prices. Local authorities could build more homes for rent and developers could erect more low-cost homes for sale. Little has yet been achieved.

## 4 Environmental issues

Speed limits for boats can limit the amount of wash caused. But to prevent erosion, speeds would have to be very low, which clashes with the main pleasure of the sport – going fast! The speed limit on Windermere is 18 kph. Limiting the noisiest and most damaging sports to certain parts of the lake can restrict the amount of damage done.

## Tourism conflicts and opportunities

### 1 Farming

Tourism and farming are often thought to be in conflict, which can be true. Visitors can trample crops and disturb livestock, but signs and education have limited these problems. Tourists have offered hill farmers new opportunities for diversification in difficult economic times. Income can be made from B&B accommodation, holiday cottages converted from farm buildings, camping and caravan sites. Activities such as pony trekking and paintballing can be offered.

### 2 Employment

The impact of tourism on employment must be positive as so many jobs are created. Many businesses thrive and make a profit. Conversely, seasonality is a problem, as well as low pay. Visitor numbers can be unpredictable.

**I** Elterwater village

## Activities

**1** Study photo **H**.

a How can overcrowding spoil people's enjoyment of Lake Windermere?

b What environmental problems are affecting Lake Windermere?

c Suggest ways in which the lake could be managed more effectively.

**2** Study photo **I**. Elterwater is a small village near Lake Windermere. This pretty settlement is close to a small lake offering flat walking, and the more demanding Great Langdale and Little Langdale walks are close by. The village has few parking spaces. Visitors park in residents' private spaces and right under their windows, making noise an issue. Outside the village, there is limited off-road parking so many cars find spaces on grass verges along the approach road.

How could the National Park Authority:

a improve the quality of life for Elterwater village residents?

b improve the recreational experience of visitors with various interests?

## ∞links

Find out more about the Lake District at www.lake-district.gov.uk.

## 8.5 Why do so many countries want mass tourism?

### Advantages and disadvantages of mass tourism

**Mass tourism** involves large numbers of tourists coming to one destination (photos **A** and **B**). There is usually a particular purpose and a particular type of location, such as skiing in a mountain resort or sunbathing at a beach location. Many countries and regions want to develop mass tourism because they believe it will bring many advantages (table **C**).

**A** Mass tourism in Barcelona, Spain

**B** The Parthenon in Athens, Greece, is the most visited ancient monument in Europe

### In this section you will learn

the term 'mass tourism'

about Jamaica – an established tropical tourist area that attracts a large number of visitors

some strategies to counter the negative environmental effects of tourism in Jamaica and to ensure its continued success.

### Did you know ??????

In the USA the first mass tourist resorts were Atlantic City in New Jersey and Long Island in New York State. In Europe, Ostend developed to serve the citizens of Brussels, Deauville attracted Parisians and Boulogne, people from Normandy.

### Key term

Mass tourism: tourism on a large scale to one country or region. This is linked to the Development and Consolidation phases of the Butler tourist resort life-cycle model.

**C**  *Advantages and disadvantages of mass tourism*

| Advantages | Disadvantages |
|---|---|
| • Tourism brings jobs. People who previously survived on subsistence agriculture or day labouring gain regular work with a more reliable wage. | • The activity may be seasonal – skiing only happens in winter. Local people may find themselves out of work for the rest of the year. |
| • New infrastructure must be put in place for tourists – airports, hotels, power supplies, roads and telecommunications. These also benefit the local population. | • The industry is dominated by large travel companies who sell package holidays by brochure or on the internet.<br>• New construction can damage the environment and cause pollution. |
| • Construction jobs often go to local people, but they are temporary. | • Lower- and middle-income customers are the target market – this type of tourism does not appeal to wealthier groups of people. |
| • New leisure facilities may be open to local people. | • Few local employees are well paid. The higher level jobs are often taken by people from the companies involved in developing the resort, who are not locals. |
| • The economy benefits from tourism through taxation, the creation of jobs and spending money in shops, hotels and at attractions. | • Investing companies are usually based In foreign countries. Profits therefore go outside the tourist country and they do not benefit the host country. |
| | • New building developments need land. Local farmers may be tempted to sell their land, which reduces local food production. |
| | • Tourists can be narrow-minded and often prefer familiar food, so much is imported rather than produced locally. |
| | • Culture clashes may occur and tourism can lead to problems with drugs and alcohol. Sex tourism is a problem in some areas. |

**Case study**

## Tourism in Jamaica

Jamaica is one of the Caribbean's main tourist destinations, with 1.9 million visitors in 2010. After this, competition from other islands began to be something of a problem. Tourism is the country's second biggest

earner and 262,000 (2011) Jamaicans work in this sector. Other local businesses also depend on tourism, such as food production for visitors and other hotel suppliers. Jamaica has much to offer the tourist (map **D** and fact file **E** on page 179) including watersports, for which it is famous, wildlife sanctuaries and, increasingly, golf.

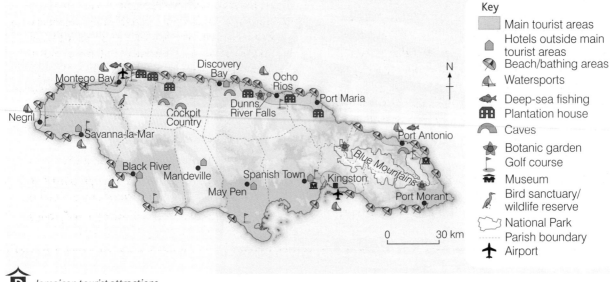

**D**  *Jamaican tourist attractions*

## Community tourism

A recent trend has been the growth in community tourism where visitors stay in local people's homes. Families provide bed and breakfast accommodation and local businesses, such as restaurants and bars, supply their other needs. This style of holiday provides greater interaction between tourists and local people, gives the visitors a clearer idea of local life, supports local businesses and uses fewer resources. Money goes directly to the people rather than to large international businesses.

**E**  *Jamaican tourism fact file, 2010*

| Jamaican tourism facts and figures | |
| --- | --- |
| Total number of tourist arrivals | 1.9 million |
| Population employed in tourism | 22.6% |
| Contribution to GDP | 24% |
| Passengers arriving by cruiseships | 910,000 |
| Rooms available to tourists | 31,000 |
| Visitors from USA and UK (% of total) | 70% and 10% |

## ∞ links

Find out more about Jamaican tourism at **www.visitjamaica.com**.

## Ecotourism

Jamaica needs to maintain its tourist resources into the future and some companies offer ecotourism, which is becoming more popular. This utilises the inland area of the island such as the Blue Mountains as well as parts of the coast, spreading tourists further around the island. Nature reserves are increasing and eco-lodges are being built. Tourist densities are kept low in these areas, which keeps pressure off the environment.

**F**  *Ocho Rios, Jamaica*

## Activities

**1**  Study map **D** on page 178 and fact file **E**.

a  Name two main tourist areas in Jamaica.

b  Which tourist activities are found in these areas?

c  Make two separate lists of the advantages and disadvantages to Jamaica of the holiday industry.

d  Research on the internet or in holiday brochures to discover any other advantages or disadvantages you could add to your lists.

e  Write a minimum of three paragraphs to discuss whether tourism is beneficial to Jamaica's people and economy. Add a clear conclusion at the end of your discussion.

**2**  This is a group exercise. At least three members of the class should be in each group. Imagine a new tourist enterprise is being set up in Jamaica, aiming to be as eco-friendly and sustainable as possible. (Read page 183 later in this chapter to understand sustainable tourism better.) You are members of the travel company (you may be from Jamaica or from a country that tourists come from – you choose) and you are preparing your proposals for the government of Jamaica and a local environmental interest group.

a  Plan an exciting programme for the clients. Name locations and the activities on offer there.

b  What type of accommodation is provided? How is it serviced?

c  How would the workforce be recruited?

d  Emphasise the advantages of your proposals and how you would aim to overcome any difficulties.

e  Compare your plans with those of other groups.

## 8.6 What attracts people to extreme environments?

### Extreme environments and activities

**Extreme environment** tourism involves dangerous landscapes often with a difficult climate, and remote places that are sparsely settled or not occupied at all. Increasing numbers of tourists are attracted to extreme environments where they can take part in adventurous activities such as rock climbing, paragliding and white-water rafting.

Extreme environments are spread across the globe and cover a wide variety of locations including mountains, deserts, rainforests, caves and ice-covered terrain (photos **A** to **D**). Adventure activities involve an element of risk and people often choose such a trip for the adrenaline rush. Examples include ice-diving in the White Sea, north Russia, with almost freezing temperatures, and travelling across the Chernobyl Zone of Alienation in Ukraine, the area devastated by nuclear contamination in 1986. In Jamaica such activities include climbing waterfalls and cliff-diving. Adventure tourism is one of the fastest-growing types of tourism in the world.

| In this section you will learn |
| --- |
| what we mean by 'extreme' tourism |
| which environments are classified as extreme |
| what is happening in Antarctica in tourism development and the likely consequences of this |
| how to assess whether particular remote landscapes should be developed for tourism on any scale. |

**A**   *Exploring the icy landscape in Greenland*

**B**   *Setting up a camp under the stars in the desert*

**C**   *Paddling in a dug out canoe in the Amazon*

**D**   *Hiking in the foothills of the Himalayas*

**Key term**

**Extreme environments:** locations with difficult environments where tourism has only recently occured due to people wanting to visit somewhere with different physical challenges.

## The target market

Adventure tourists look for physical challenge and risks. They are often around 30 years old, unmarried and without children, have high-powered jobs and a good income – these trips are expensive. Groups are small and distances great. However, there are enough wealthy individuals with a taste for something completely different to allow this sector to grow. It will never be large but in some areas it is increasing in significance. Most companies advertise on the internet rather than by brochure.

Little investment is needed to set up such trips. The usual costly expenses of building hotels and roads are irrelevant. Part of the experience is to sleep 'rough' and travel over untouched landscapes. This tourism sector is growing rapidly in Peru, Chile, Argentina, Azerbaijan and Pakistan. Northern Pakistan is one of the most mountainous and difficult landscapes in the world and even its risky political situation as the base of Al Qaeda terrorists adds a thrill for some.

## Antarctica

In the 1950s small-scale tourism started in Antarctica when commercial shipping began to take a few passengers. The first specially designed cruise ship made its first voyage in 1969. Some 9,000 tourists in 1992–93 have now grown to 33,824 (2010–11) (table **E**). This is thousands more than the scientific workers and their support staff who are there temporarily for research purposes.

Tourists from the northern hemisphere usually fly to New Zealand or Argentina, taking their cruise ship onwards for one to two weeks. Smaller boats take them ashore at key locations for short visits, mainly to the peninsula or nearby islands (map **F**).

Landing sites on Antarctica are limited and quickly become honeypot sites. Tourists have the potential to cause environmental and ecological damage, so management is important. Walking, kayaking, skiing, climbing, scuba diving and helicopter/small aircraft flights are some of the activities offered to tourists (chart **H** opposite).

**F**   *The Antarctic Peninsula and nearby islands*

**E**   *Antarctic tourist numbers (2010–11)*

| Country of origin | Numbers | Percentage of total |
|---|---|---|
| USA | 12,629 | 37.4 |
| Australia | 3,220 | 9.5 |
| UK | 2,763 | 8.1 |
| Canada | 2,531 | 7.5 |
| Germany | 2,378 | 7.0 |
| Japan | 936 | 2.8 |
| Netherlands | 889 | 2.6 |
| Others | 8,478 | 25.1 |
| Total | 33,824 | 100.0 |

**G**   *Cruise ship in Antarctic waters*

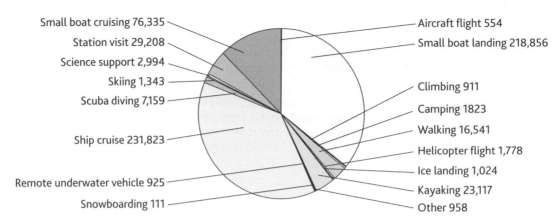

Small boat cruising 76,335
Station visit 29,208
Science support 2,994
Skiing 1,343
Scuba diving 7,159
Ship cruise 231,823
Remote underwater vehicle 925
Snowboarding 111

Aircraft flight 554
Small boat landing 218,856
Climbing 911
Camping 1823
Walking 16,541
Helicopter flight 1,778
Ice landing 1,024
Kayaking 23,117
Other 958

**H** *Tourist numbers in Antarctica, 2007–08*

Tourists only spend a short time ashore, but their impacts can be considerable. Animals, especially penguins and seals, are disturbed by more than a few people (photo **I**). Not used to humans, they do not like to be touched. If they leave as a result, they may abandon eggs and young.

There have been accidents when ships have struck uncharted rocks or ice floes. The great majority of shipping in Antarctic waters is tourist-based. Oil spills are becoming an increasing hazard for wildlife. Tourist ships must discharge all waste materials well away from the shore of Antarctica.

## Coping with tourism in Antarctica

All tour operators are members of IAATO, which directs tourism to be safe and environmentally friendly. Around 100 companies are involved. In line with the Antarctic Treaty, tourism is an acceptable activity in Antarctica – it is the scale that has to be controlled. Visitors are not allowed to visit Sites of Special Scientific Interest (SSSIs) in order to conserve precious wildlife and landscapes. Bird Island on South Georgia is one example.

Although tourist numbers have increased rapidly in Antarctica, protection remains a high priority. A permit must be gained for any activities on the continent. No ship carrying over 500 passengers can land in Antarctica. Nevertheless, there is concern that larger ships will eventually be allowed to land and that the volume of tourists will be beyond sustainable limits.

**I** *The impact of tourists can be great*

**Study tip**

Ensure that you can describe *and* explain some of the advantages and disadvantages of mass tourism to the economy and environment of a location.

**Activity**

Use the information about Antarctica to help you make a study of tourism in this extreme environment. Use the internet to help you investigate the following themes:

a What are the attractions to tourists?

b Where do tourists come from? (Here you could use table **E** on page 181 to make a flow map.)

c What are the issues associated with tourism?

d How should tourism be controlled in the future?

**Did you know** ??????

The ice in an Antarctic iceberg is around 100,000 years old. The continent is actually a desert, with only 254 mm of precipitation each year.

⦾ **links**

Find out more about tourism in Antarctica at **www.coolantarctic.com**.

# 8.7    How can tourism become more sustainable?

## ■ The aims of ecotourism

**Ecotourism** is environmentally friendly tourism. It caters for a small but growing niche market of environmentally aware tourists and is the fastest expanding tourism sector.

Ecotourists want to experience the natural environment directly, undertaking activities such as trekking and bird-watching. They want their holiday to have as little impact on the environment as possible. Energy use should be sustainable and no waste should be generated that cannot be dealt with efficiently. Ecotourists prefer small-scale accommodation in lodges that may not even have electricity, rather than large hotels. They eat local food. Local people are their guides as their knowledge and experience is seen to be more valuable. The impact on the environment is low but, because ecotourism is small in scale, the price paid by each tourist is high. The market for such tourism is therefore limited.

**A**    Ecotourism in a Costa Rican rainforest

### Case study

## The Galapagos Islands

The 50 volcanic Galapagos Islands lie 1,000 km off the west coast of South America in the Pacific Ocean. They belong to Ecuador. Here, Charles Darwin formulated his theory of evolution. Around 90 per cent of these islands are designated as National Park or marine reserve. Protection began in the 1930s. The islands are among the most fragile and precious ecosystems in the world, becoming the first Unesco World Heritage Site in 1979. The area is also a biosphere reserve and whale sanctuary.

Today, tourists visit under strict rules. They arrive mainly by small ships that tour the islands and allow people onshore only at specific locations in limited numbers.

The tour boats are owned by locals and take 10 to 16 tourists each, many accompanied by professional guides. Visitors are given accurate information and prevented from causing damage.

### Benefits

■ Environmental – the Galapagos Conservation Trust receives approximately £25 from each tourist out of the cost of their holiday. This pays for conservation work on the islands.

■ Economic – local people act as paid guides and small businesses have been started, for example boat companies providing trips around the islands. Local people run guest houses and provide other services.

### Problems

■ Honeypot sites – even though tourism is controlled, some sites are over-used and are showing signs of environmental stress.

■ Pollution – oil pollution from boats can affect fragile marine ecosystems.

### Activities

**1**  **a**  Write a definition of ecotourism.

   **b**  What are the main characteristics of ecotourism?

   **c**  Do you think ecotourism is a good thing? Why?

**2**  The Geography and Biology departments at your school want to run a trip to the Galapagos Islands. Your class has been asked to help promote the trip by producing posters and PowerPoint presentations. The trip is to be as environmentally sensitive as possible so you need to stress the principles of ecotourism. Work in pairs or small groups and use the Internet to help you.

### Study tip

Ensure you can use an example to explain how ecotourism can contribute to sustainable development.

# 9 Geographical skills

The purpose of this chapter is to provide you with a summary of the Geographical skills identified in the specification checklist. The appropriate methods and techniques are described and, where appropriate, there are some activities to help you to understand and apply them. Two full page map extracts, at 1:50,000 and 1:25,000, are included together with a more comprehensive set of activities to give you useful practice.

## ■ Maps and photos

### Four-figure and six-figure grid references

Ordnance Survey (OS) maps have gridlines drawn on them to enable locations to be given. The lines that run 'up and down', and increase in value from left to right (west to east), are called eastings. Those that run across the map, and increase in value from bottom to top (south to north), are called northings.

To locate a grid square on a map, we use a four-figure reference. The first two digits refer to the easting value and the second two digits to the northing value. The four-figure reference for Square A on Figure **A** is 3478; notice that Square A is the square after the values 34 and 78.

To locate a point rather than a square, each square is split into tenths to give a six-figure reference. Look at Figure **A** and notice that Point X is at grid reference 355762. Notice how the eastings value is represented by the three digits 355 and the northings value is represented by the digits 762. It is the third digit of each set that is the 'tenths' value. Thus, the eastings value is 35 and 5/10ths and the northings value is 76 and 2/10ths.

The key thing with grid references is to remember to give the eastings value first and then the northings value. Some people use the phrase 'along the corridor and up the stairs' to get the order right!

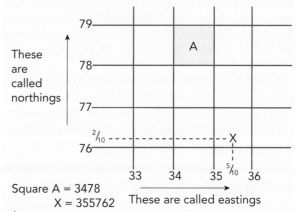

Square A = 3478
X = 355762

**A** *How to find grid references*

### Compass directions

Most of the time on a map, the direction 'north' is 'straight up', but it is very important that you check the key when examining maps and diagrams. The Certificate course requires you to know the eight points of the compass (see Figure **B**).

Be sure to express a compass direction carefully and precisely, for example, 'Settlement X is to the north-west of Settlement Y'.

**B** *The eight points of the compass*

# Drawing cross-sections

A cross-section is an imaginary slice through a landscape. It is very helpful in being able to visualise what a landscape actually looks like, and it is one of the most important mapwork skills employed by a geographer. Make sure that you can identify and label the main physical features of a landscape, for example, steep and gentle slopes, ridge, escarpment and valley.

To draw a cross-section you need a piece of paper, a sharp pencil, a ruler and a rubber. The stages of construction are shown in Figure **C**.

As you complete your section, bear in mind the following points:

- double-check your accuracy of copying height values

- try to make your vertical scale as realistic as possible; don't exaggerate it so much that you create a totally unreal landscape!

- complete the section to both vertical axes by carrying on the trend of the landscape

- label any features

- complete axes labels and give grid references for each end of your section

- give your section a title.

## Measuring straight and curved distances

Distance can be measured as a 'straight-line' distance or as a 'curved' distance, say along a road or a river.

Every map will have a scale, usually in the form of a measured line (called a linear scale) with distances written alongside. To calculate a straight-line distance, simply measure the distance on the map between the two points in question, using a ruler or the straight edge of a piece of paper.

**A**

- Place the straight edge of a piece of paper alongside the chosen line of section.
- Mark the start and finish of your section.
- Mark the contours and features, e.g. rivers.

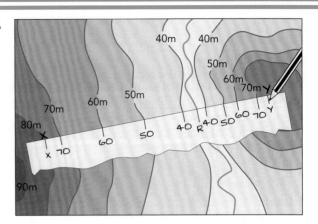

**B**

- Draw axes of a graph. Take care to choose an appropriate vertical scale.
- Lay your paper along the horizontal axis and locate each contour value on the graph paper, marking each with a cross.

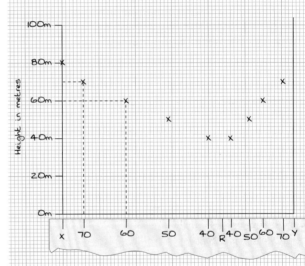

**C**

- Join up the crosses with a freehand curve.
- Label any features.
- Complete the graph with a title.

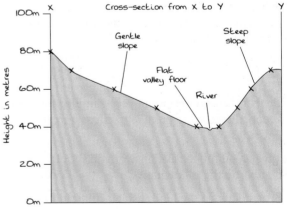

 **C** *Drawing cross-sections*

You then line up your ruler or paper alongside the linear scale to discover the actual distance on the ground in kilometres.

A curved distance takes rather longer to work out. The best technique is to use the straight edge of a piece of paper to mark off sections of the curved line, effectively converting the curved distance into a straight-line distance. Look at Figure **D** to see how this technique works.

**Straight-line distance**

1 Use a ruler to measure the distance between two places on the map, in centimetres.

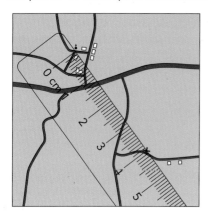

2 Measure out the distance on the map's linear scale to discover the distance on the ground in kilometres.

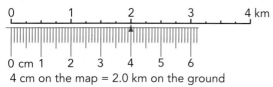

4 cm on the map = 2.0 km on the ground

**Curved-line distance**

1 Place the straight edge of a piece of paper along the route to be measured. Mark the start with the letter S. Look along the paper and mark off the point where the route moves away from the straight edge.

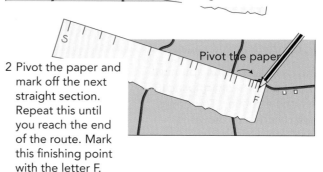

2 Pivot the paper and mark off the next straight section. Repeat this until you reach the end of the route. Mark this finishing point with the letter F.

3 Place the edge of the marked paper alongside the linear scale on the map and convert the total length to kilometres. Remember to always give the units when writing your answer!

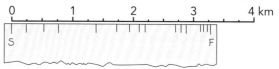

**D** *Measuring distance*

## Interpreting maps

### Relief

Relief is the geographical term used to describe the lie of the land. Exam mapwork questions frequently ask for descriptions of relief. It is important that you refer to simple landforms such river valleys, hills and escarpments and that you use adjectives to develop your description. For example, 'there is a steeply dissected river valley with asymmetrical valley sides'.

To give a comprehensive answer you should comment on the following:

- The height of the land, using actual figures taken from contours or spot heights to support your points. Using words like 'high' and 'low' are fairly meaningless without the use of actual figures.
- The slope of the land – is the land flat, or sloping? Which way do the slopes face? Are the slopes gentle or steep? Are there exposed, bare cliffs? Again, it is important to give precise supporting information such as grid references and compass directions.
- The presence of features such as valleys, dry valleys and escarpments. Refer to names and use grid references.

### Drainage

Drainage is all about the presence (or absence) and flow of water. When describing the drainage of an area, try to comment on the following:

- The presence or absence of rivers. Which way are they flowing? (Look at the contours.) Are the rivers single or multi-channelled? Give names of the rivers, and use distances, heights and directions to add depth to your description.
- Drainage density – this is the total length of rivers in an area, usually expressed as 'km per square km'. High drainage densities are typically found on impermeable rocks, whereas low densities suggest permeable rocks.

- The pattern of rivers (see Figure **E**).
- The influence of people on drainage channels, for example, straightening channels, building embankments, etc. Straight channels are rare in nature and usually indicate human intervention.
- Evidence of underground drainage, in the form of springs or wells.
- Presence of lakes, artificial or man-made.

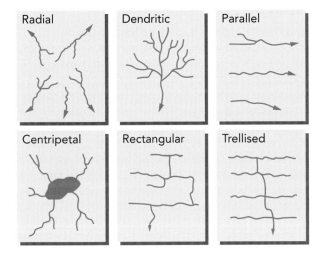

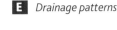
**E**  *Drainage patterns*

## Settlement

When describing the patterns and types of settlement, you should understand the following geographical terms (Figure **F**) :

- Dispersed settlements – these low density settlements are spread out over a large area and are typical of rural agricultural regions.
- Nucleated settlements – these are high density settlements, tightly packed and often focused on a central point such as a major road intersection. The built-up area typically spreads out in all directions unless there is a physical constraint such as a coastline or mountain range.
- Linear settlements – these are settlements that typically extend along a road, railway or canal. 'Linear' means 'line' so on an OS map a linear settlement tends to be long but narrow in its extent.

| Dispersed | Nucleated | Linear |
|---|---|---|

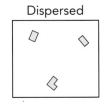 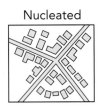 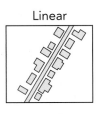

**F**  *Settlement types*

## Communication

Communication networks include roads (of various types), railways and footpaths, as well as less common forms of transport such as cross-channel and river ferries, airports and cycle ways.

You should be able to describe these networks, giving locational details such as length and orientation or compass direction and referring to patterns and density. For example, roads may radiate out from a settlement or form a series of concentric ring roads and by-passes.

Communication networks frequently reflect the relief of an area. Major transport arteries such as roads, canals and railways tend to follow flat, low ground, which explains why they are often located in river valleys. Footpaths also tend to follow river valleys, as well as linking settlements and following ridge-lines or escarpments. Look out for named footpaths, such as the Pennine Way, and make sure that you refer to them by name in an exam answer.

## Land use

As you would expect, land use refers to the way the land is used! You need to make full use of the key when considering, for example, different types of woodland (for example, coniferous or non-coniferous), coastal deposits (mud, sand or shingle) and vegetation (for example, scrub, bracken or marsh). Other land uses include urban areas (be prepared to describe settlement patterns), fields (often just shown white on maps), quarries, industrial areas and tourist sites. Indeed, land use includes all aspects of the Earth's surface!

Be prepared to 'describe' land use by referring to specific locations (don't forget to use grid references), size, area and shape. Also be prepared to 'explain' why certain land uses exist, for example, why there are no rivers on a limestone plateau or why there are rivers and reservoirs on Dartmoor (granite, an impermeable rock). The reasons why certain land uses are present often form the trigger for a higher level mark in an exam question. So, try to use the word 'because' when you are asked to explain a land use.

## Identify and describe basic landscape features

Patterns of contours on maps can be used to identify basic physical features, such as river valleys, ridges and plateaus (see Figure **G**). Having identified these features, you need to be prepared to describe them by referring to size, shape, heights and orientations.

For example:

'The asymmetrical ridge is about 2 km wide, is orientated roughly north-south and rises to a maximum height of 232 m at GR 376490. It has a steep eastern side and gentle western side.'

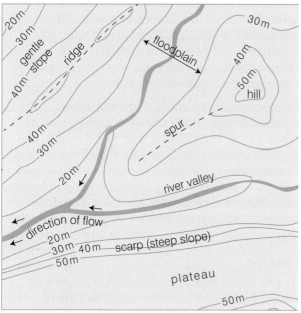

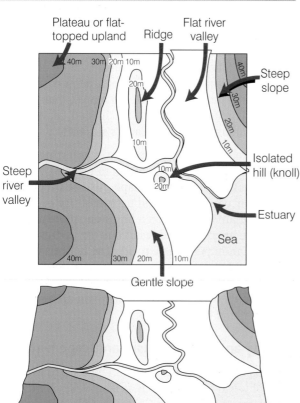

## Describe physical features of coastlines, shape and form of river valleys

Whether or not you are studying the units on coasts and rivers, you do need to have a basic understanding of the key physical features associated with coastlines and river valleys. This requires both identification and description, so be prepared to refer to size, shape, orientation, etc.

### Coastlines

Whilst some stretches of coastline are largely straight (usually reflecting the presence of a single rock type), many coastlines are indented with headlands and bays. Headlands often exhibit features of coastal erosion, such as cliffs and stacks (isolated pillars of rock). Bays tend to exhibit features of coastal deposition, such as beaches; sand blown onshore can form sand dunes too. Sediment carried along a coastline by the process of longshore drift will form narrow fingers of land called spits, commonly found in river estuaries or where the shape of the coastline changes suddenly.

These features are described in detail in Chapter 4.

### River valleys

As a river flows from source to mouth its valley usually becomes wider and less steep-sided. In the uplands, close to its source, a river will flow through a steeply dissected V-shaped valley, often around spurs of rock (called interlocking spurs) that jut down into the valley. Further downstream, as the river starts to erode laterally as well as vertically, the valley becomes wider and the river develops a broad, flat floodplain. The edge of the floodplain, where it comes into contact with the river valley, is marked by a break of slope called a bluff.

These features are described in detail in Chapter 3.

## Interpret and describe features of urban morphology, settlement functions and services and infer human activity

As a geographer, you will be expected to describe and interpret features on a map extract as well as simply identifying them. In addition to the physical features referred to previously, there are a number of human features too, most particularly those associated with settlements.

### Urban morphology

The word 'morphology' describes the shape and characteristics of particular features, in this case settlements.

- CBD – at the heart of towns and cities is a central zone characterised by shops, offices and some residential and administrative buildings. This zone is the central business district, or CBD. The CBD often contains historic buildings, such as a cathedral or a town hall, and bus and train stations are also commonly located here.

- Inner city – this is the zone adjacent to the CBD, often characterised by a mixture of older working-class homes (which may have been renovated and modernised) together with evidence of changing land use, for example, where old terraced housing has been replaced with modern offices or tower blocks. There may be some derelict areas and open spaces awaiting redevelopment. Inner cities usually have a variety of land uses and reflect the changing economic conditions of a town or city.

- Suburbs – suburban housing, together with shopping parades, is typically found at the outskirts of towns and cities. There will often be several types of housing, ranging from rows of semi-detached properties built in the 1930s or 1950s to modern housing estates with cul-de-sacs and a mixture of bungalows, detached and semi-detached homes. The main characteristic of the suburbs is the dominance of residential land uses.

- Rural–urban fringe – this is the area on the edge of a town or city where it merges into the surrounding rural countryside. There will be some urban land uses, such as ring roads, out-of-town retail parks and modern industrial estates and other land uses that require plenty of space, such as hospitals, schools and cemeteries. Golf courses are often found in the rural–urban fringe along with some more typical rural land uses such as keeping horses in fields.

## Settlement functions and services

Settlement functions and services describe what happens in a particular town or city and, essentially, what makes it tick! Most settlements provide sources of employment, housing, retailing, health care, leisure and recreation. Some will be administrative centres, for example, county towns like Cambridge or Worcester. Some will have developed a particular function as a resort (for example, Blackpool), a university town (for example, Oxford), a port (for example, Bristol) or an industrial town (for example, Middlesbrough).

When using a map or a photo, you need to look for evidence and then suggest (infer) what type of settlement it is or what type of urban zone you are looking at. Inference is all about reaching informed conclusions using the evidence available to you. So, if you are asked to identify the 'inner city', look for the appropriate evidence on the map or photo and then use it to support your suggestion.

Figure **H** identifies some of the key characteristics of the small Northumberland market town of Alnwick.

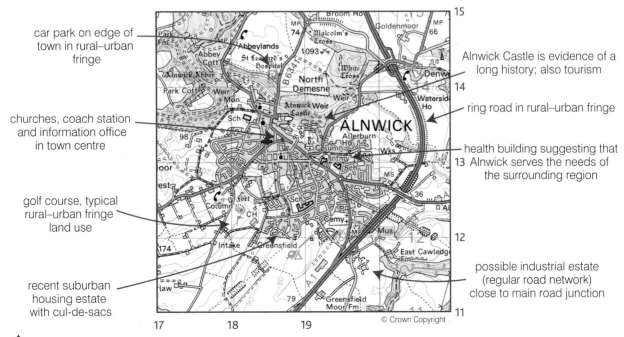

car park on edge of town in rural–urban fringe

churches, coach station and information office in town centre

golf course, typical rural–urban fringe land use

recent suburban housing estate with cul-de-sacs

Alnwick Castle is evidence of a long history; also tourism

ring road in rural–urban fringe

health building suggesting that Alnwick serves the needs of the surrounding region

possible industrial estate (regular road network) close to main road junction

© Crown Copyright

**H**   *Characteristics of a small market town: Alnwick (a nucleated settlement)*

## Use and interpretation of ground, aerial and satellite photos to describe landforms, natural vegetation, land use and settlement

Photos are widely used by geographers to record, investigate and understand physical and human features. They are also commonly used in exams, which is why you need to feel confident about their use. You will find examples of ground, aerial and satellite photos in this book and will use them throughout your course.

When describing what a photo shows, make sure you use directional language, for example, 'in the foreground', 'in the background' as well as 'right' and 'left'. Use juxtaposition (for example, 'just behind and to the right of the stack') to enable you to identify and describe features accurately.

It is quite likely in an exam that you will be asked to use photos and maps alongside each other. You could, for example, be asked to identify the direction that a photo is looking. Take your time to orientate the photo on the map, possibly by turning it to line it up correctly – look for evidence in the foreground and the background on the photo to help you do this. Once you have orientated the photo, it is relatively straightforward to work with both resources.

- Ground photos – as the name implies, these photos are taken on the ground! They are the most common types of photo and are usually used to focus on a particular feature or characteristic, such as a building or part of a cliff. Look at the ground photos on page 55 to see how they are used to show the features of Grindsbrook Clough and the River Noe.

- Aerial photos – taken by aeroplanes or helicopters, aerial photos look down on a landscape often showing large areas that can be related directly to OS maps, for example, showing settlements or stretches of coastline. Vertical aerial photos look directly down on to the ground and therefore give no indication of relative height; everything looks flat! Oblique aerial photos give a sideways view of the landscape. They tend to be used more commonly than vertical aerial photos but they do distort size, with objects in the foreground appearing much larger than those in the distant background. Look at the oblique aerial photograph of the Cockermouth flood on page 68.

- Satellite photos – you will be familiar with satellite photos if you have accessed Google Earth or even just Google Maps. In common with vertical aerial photos, they look directly down on to the Earth. Satellite photos can either be straightforward unaltered photos or they can be digitally processed and enhanced to use colours to make certain land uses and features show up more clearly. These 'false-colour' images have been widely used to show environmental concerns such as pollution and deforestation. One of the best ways for you to familiarise yourself with satellite photos is to use Google Maps and tab between 'map' and 'satellite' for an area that you know well. Satellite photos are commonly used to show weather features, such as depressions and hurricanes.

## Drawing sketch maps

A sketch map is drawn to provide a simplification of an OS map. It enables the drawer to focus on just a few key elements, such as patterns of roads or rivers, without the confusion of lots of other unnecessary and potentially confusing information. Drawing a sketch map is an important geographical skill and one that you could be asked to demonstrate in an exam.

- Start by drawing a frame, either to the same scale as the map or enlarging/reducing it as required.

- Divide the frame into grid squares as they appear on the map and write the numbers at the edges of your frame. These will act as your guidelines when you draw your sketch.

- Using a pencil, carefully draw the features that you have been asked to draw or that are relevant to the task in hand. Do not copy every detail on to your sketch.

- Once complete, you can use colour and shading if you wish, although black and white sketches are often the most successful.

- Label and annotate as required, and don't forget to include a scale (which may be approximate), a north point and a title.

- Remember that most of the credit for a sketch map will relate to your labels and annotations (detailed labels often with some explanation), which show your ability to interpret the map.

## Drawing sketches from photos

It is important to realise that the purpose of a sketch is to identify the main geographical characteristics of the landscape (Figure I). It is not necessary to produce a brilliant artistic drawing; clarity and accuracy are all that are needed so that labels and annotations can be added.

To draw a sketch, you first need to draw a frame to the same general shape of the photograph. Then draw one or two major lines that will subsequently act as guidelines for the rest of your sketch. You could draw the profile of a slope or a hilltop, or a road or river, for example. Consider what it is that you are trying to show and concentrate on these aspects; it may be river features or the pattern of settlements. Don't take time drawing a lot of detail that is not required and only serves to confuse.

Always use a good sharp pencil and don't be afraid to rub things out as you go along.

Finally, remember to label or annotate (detailed labels) your sketch, to identify the features, and give your sketch a title.

## ■ Atlas maps

Atlas maps are extremely useful sources of information for geographers. They provide a wealth of information:

- Basic maps of countries and regions of the world showing physical relief, settlements and political information.
- Thematic maps, such as those showing climate, vegetation, population, tourism and tectonics.
- Maps associated with global issues, such as pollution, global warming, desertification and poverty.
- Tables of statistics.

You will make use of a range of atlas resources while you are studying Geography and it is possible that you might have an atlas map in your exam. Exam questions tend to focus on patterns or distribution, so be sure that you practise these skills.

A pattern implies a degree of regularity or connection, say by population being concentrated along the coast or industry being concentrated along a river valley. Terms such as 'radial' (spreading outwards from a central point) or 'linear' can be used to describe a pattern. For example, earthquakes tend to form a linear pattern in the North Atlantic by following the North American/Eurasian plate margin.

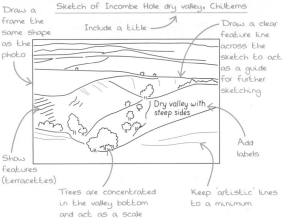

**I** *Sketch from a photo*

'Distribution' is a term that is used more broadly to describe where things are; there may or may not be a regular pattern. Take the distribution of population in Kenya, for example. Most people live in the highlands, where the climate is less extreme and the soils provide good land for farming. Fewer people live on the lower ground in Eastern Kenya because it is very hot and dry and much less suitable for habitation.

Examples of the atlas maps that you can expect to see in an exam include population distribution, population movements, transport networks, settlement layout, relief and drainage. Be prepared to make use of two maps to consider links and relationships, for example, between population distribution and relief in Kenya.

# 10 Graphical skills

As a geographer, you need to be able to read and interpret information that is presented in a variety of ways. This includes written text, photos and maps of different types and scale. It also includes information presented in the form of graphs and diagrams. Not only do you need to be able to interpret information, but you need to be able to construct a variety of basic graphs and diagrams.

## How to interpret a graph or diagram

Look at Figure **A**. It is a line graph showing world population growth.

- Notice that the scales have equal intervals between each line (2 billion people on the *y* axis and every 50 years on the *x* axis). If different intervals were used, the graph would be distorted.

- The top line of the graph shows the total world population. You can see that it grows slowly from about 1 billion in 1800 to about 3 billion in 1950. From 1950 it grows very rapidly before levelling off at about 10 billion by 2050.

- The graph has been sub-divided into Developing regions and Developed regions. It can be called a 'divided' or 'compound' graph. You need to exercise care when giving population figures for each part of the graph by using either the green shaded area or the yellow shaded area. So, if asked for the Developing regions population in 1950, it is about 2 billion.

Remember the three stages of a description:

1 Describe the overall trends and patterns – the 'big picture'.
2 Provide some evidence to support your description (quote a few facts and figures).
3 Consider any anomalies (exceptions) to the overall trends and patterns.

When explaining:

- try to give some reasons for the trends/patterns/anomalies (use the 'because' word)

- consider links and connections between different variables that might help your explanation (for example, a war might affect the shape of a population pyramid).

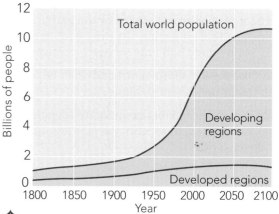

**A** *World population growth line graph*

## How to construct graphs and diagrams

It is possible that in an exam you will be asked to construct or, more likely, complete a partly drawn graph, diagram or map. This should be fairly straightforward as long as you look closely at the scales and plot the data accurately and with precision – use a sharp pencil and double-check that you have plotted the information correctly.

It is surprisingly common for candidates to overlook these practical activities in an exam and lose marks just because there are no lines for writing an answer, so be vigilant!

### Line graphs

A line graph shows continuous changes over a period of time, for example, stream flow (hydrograph), traffic flow or population change. It is a very common and effective technique to use, but the most important thing to remember is that time, shown on the horizontal axis, must have an equal spacing, for example for periods of time (Figure **A**).

## Bar graphs and histograms

Bar graphs and histograms are two of the most common methods used to display statistical information. However, there are differences between them and you should try to use the technique that is best suited to your data.

A bar graph provides a visual display for comparing quantities or frequencies in different categories, such as types of vegetation or places. The bars are drawn with a gap in-between them and coloured or shaded differently because they are unconnected (Figure **B**).

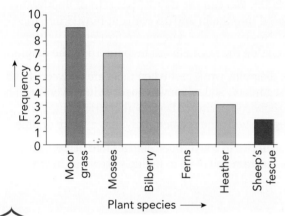

*Bar graph to show the frequency of plant species in a vegetation quadrat sample*

A histogram also uses blocks but with no gaps in-between. This is because a histogram is drawn when there is continuous data (such as daily rainfall values over a period of a month) or the values are all part of a single survey, for example, the sizes of particles in a sediment sample (Figure **C**). As the bars are effectively connected, a single colour or type of shading is used.

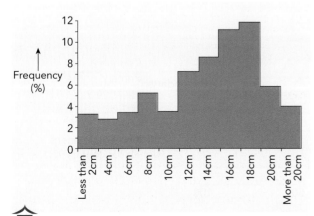

**C** *Histogram to show size distribution of beach pebble*

It is possible to sub-divide individual bars in order to show multiple data. Look at Figure **D** and notice that each bar has been sub-divided to show different types of crop. This type of graph is called a divided bar chart.

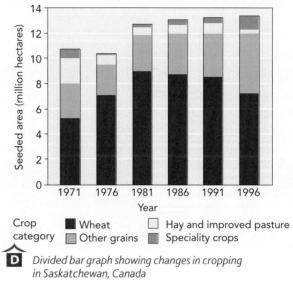

**D** *Divided bar graph showing changes in cropping in Saskatchewan, Canada*

## Scattergraphs

If two sets of data are thought to be related, they can be plotted on a graph called a scattergraph. To complete a scattergraph:

- draw two axes in the normal way, but try to put the variable thought to be causing the change (called the independent variable) on the horizontal ($x$) axis. In Figure **E**, GNP (Gross National Product, the wealth of a country), is thought to be responsible for average car ownership.
- use each pair of values to plot a single point on the graph using a cross
- use a best-fit line to clarify the trend of the points if there is one. Your best-fit line should pass approximately through the centre of the points so that there is roughly the same number of points

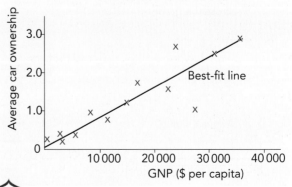

**E** *Scattergraph and best-fit line*

on either side of the line. Use a ruler to draw a straight line. The best-fit line does not need to pass through the origin.

■ describe the resultant pattern (see Figure **F**).

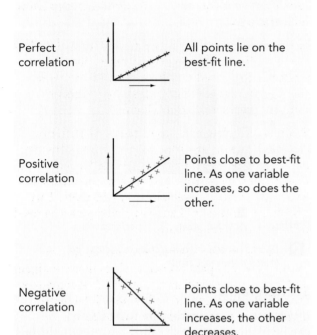

Perfect correlation — All points lie on the best-fit line.

Positive correlation — Points close to best-fit line. As one variable increases, so does the other.

Negative correlation — Points close to best-fit line. As one variable increases, the other decreases.

Poor correlation — Points suggest no clear pattern.

**F** *Interpreting scattergraph patterns*

## Triangular graphs

A triangular graph enables three values to be plotted at the same time to produce a single point. The values take the form of percentages that add up to 100 per cent. They are commonly used to show soil texture, employment and types of mass movement. Figure **G** explains how to plot a value on a triangular graph.

## Pie diagrams

A pie diagram is quite simply a circle divided into segments, rather like the slices of a cake. It is usually drawn to show the proportions of a total. For example, Figure **H** shows immigration into the EU in 2006. You can see that 27 per cent of immigrants moved into Spain and 15 per cent moved into the UK. Notice that the percentage figures are written alongside the segments to help interpret the diagram.

Pie diagrams work best when there are between 4 and 6 'segments', and when you are drawing several graphs to show comparisons. Don't bother drawing a pie diagram with one segment only; it is a waste of time.

Remember, to convert percentages into degrees for your pie diagram you need to multiply the value by 3.6.

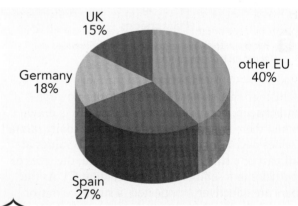
**H** *Pie diagram showing immigration into the EU, 2006*

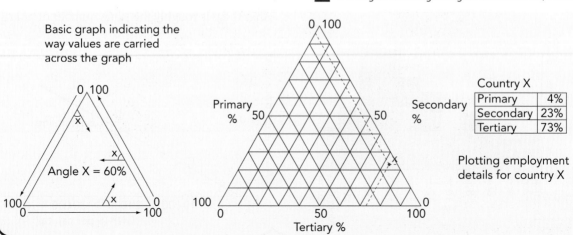
**G** *Triangular graph*

## Choropleth maps

A choropleth (note that it is not spelt chlropleth!) map uses different colours or density of shading to show the distribution of data categories. Figure **I** shows a choropleth map. Notice the following key features:

- The base map shows regions or areas, in this case countries of the world.
- Data is divided into a number of groups or categories. Ideally there should be between 4–6 categories. Notice that the category values do not overlap and that the intervals are equals (for example, 2–2.9, 3–3.9, etc).
- The darker (or denser) the shading, the higher the values. This is a key characteristic of a choropleth map as it creates a powerful and immediate visual impression. The sequence shown in Figure **I** (yellow – orange – red – brown) works well. A black could be added after brown. White (blank) can be used for the lowest category as long as there are no areas on the map for which no data is available.

Choropleth maps are very effective means of showing variations that exist between areas. However, they can be misleading in that there will often be significant variations at a local scale within each area. They also imply sudden changes at the boundaries of the areas, which is often not the case on the ground.

## Isoline maps

An isoline map is a map that uses lines of equal value to show patterns ('iso' means 'equal'). Some of the most common types of isoline map are drawn to show aspects of weather and climate, for example, isobars show pressure and isotherms show temperature.

Isoline maps are rather tricky maps to draw but they are very effective at showing patterns, particularly when superimposed on a base map. For example, they are good at showing pedestrian counts at different places in a town (Figure **J**).

To draw an isoline map, you need to mark your observed data on to a base map or sheet of tracing paper/acetate. You then need to consider how many lines to attempt to draw and at what intervals you will draw them, for example, every 10 units. This decision is largely 'trial and error' and you may need to have a go in rough first.

Look at Figure **K** to see how isolines are drawn. Notice how they pass between values that are higher and lower than the value of the line. Just remember that all values to one side of a line will be higher and all those to the other side will be lower.

There is a degree of individual determination and decision making, so do not worry if your map turns out to be slightly different from your neighbours.

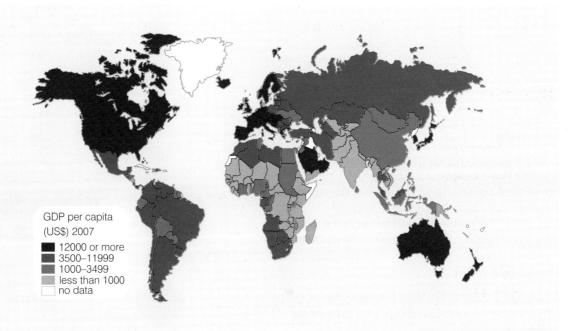

GDP per capita (US$) 2007

- 12000 or more
- 3500–11999
- 1000–3499
- less than 1000
- no data

**I** *Choropleth map showing GDP per capita, 2007*

**J**   *Isoline map showing pedestrian counts in Blackburn's CBD*

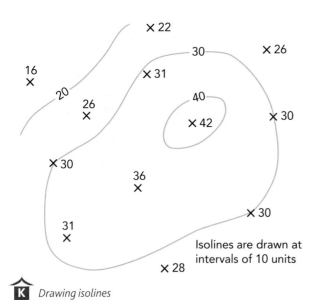

Isolines are drawn at intervals of 10 units

**K**   *Drawing isolines*

## Dot maps

Dots are used to represent a particular value or number (for example, a population of 1 million) and are located accurately on a map. It is the density of the dots that creates the visual impact of the map. Dot maps give powerful impressions but it is not easy to extract accurate information from them.

## Desire line maps

Desire lines show movement between places, for example, commuters travelling from their home villages to a nearby town or city. They show trends in the distance travelled between places (for example, most people might commute short distances) and the spatial density of the travellers themselves (for example, most people might migrate from a single village in the south).

When drawing a desire line map, you just need to be accurate in positioning each line to show its source and its destination.

## Proportional symbols

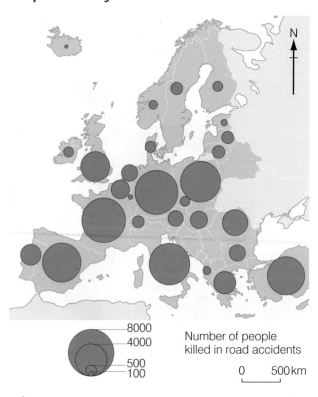

Number of people killed in road accidents

8000
4000
500
100

0   500 km

**L**   *People killed in road accidents in Europe, 2000*

Proportional symbols (for example, circles) are very effective ways to show data, particularly on a base map where spatial variations can be seen. They are, however, rather tricky to draw and you will need to choose your scale carefully.

In drawing proportional circles, select a scale for the radius of your circle. As it is the area of the circle that needs to be proportional, you have to use the square root value as your radius distance (see Figure **L**).

## Flow line maps

Flow lines are similar to desire lines in that they show movement between places by connecting the source with the destination. However, whereas desire lines show lots of individual movements, flow lines are drawn proportionately to show grouped data.

For example, 10 separate desire lines might be drawn to show the movement of commuters from Village A to Town B. An alternative would involve drawing a single flow line whose width was drawn proportionately to show the 10 commuters.

Look at Figure **M**. It shows the origin of tourists visiting Kenya. Notice that each flow line connects the continent of origin and the destination (Kenya) and that each one has been drawn with its width in proportion to the percentage of tourists (1 mm = 2 per cent total tourists).

Flow lines are particularly effective when drawn on a base map, but time and care are needed to choose an appropriate scale to avoid flow lines crossing over each other. Don't forget to write the scale on your map.

## Topological maps

A topological map differs from an ordinary map in that it distorts true spatial distance or area. One of the best known examples of a topological map shows the London Underground. It accurately shows links and connections between stations, but is not drawn to be an exact representation of the real world.

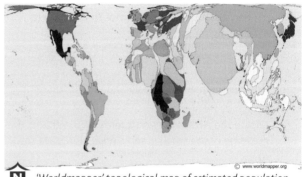

**N**   'Worldmapper' topological map of estimated population, 2050

Look at Figure **N**. This map distorts the true area of countries in an attempt to represent a different factor, in this case estimated population in 2050. Each country is drawn as a proportion of the total global estimated population. Topological maps like this convey important messages and are often constructed to make a point.

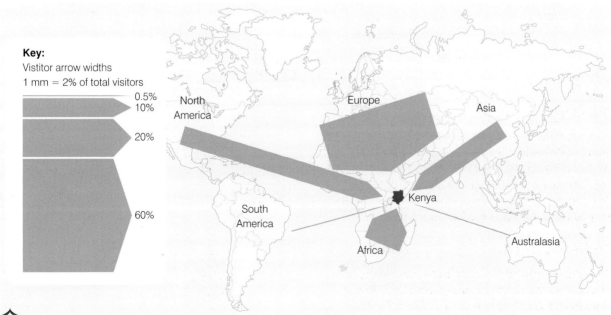

**M**   Flow line map showing the origin of tourists to Kenya

# 11 Map activities

## Activities

1:50,000 Ordnance Survey map extract: Amble (see page 202 for key to symbols)

**1**

a Locate the village of Alnmouth and give its four-figure grid reference.

b What type of woodland is found in grid square 2402?

c What is the name of the village in grid square 2401?

d Locate the A1068 in grid square 2408. What is the meaning of the small green circles alongside the road?

e Give the six-figure grid reference of the Information Centre in Amble.

f Give the six-figure grid reference of Alnmouth train station.

g What is found at 232066?

h What is found at 258037?

**2** Read through the following sentences that use compass directions. Decide which are true and which are false.

a Warkworth is to the north-west of Amble.

b Alnmouth is to the south-west of Lesbury.

c If you go from Alnmouth to Warkworth, you would be travelling in a southerly direction.

d Acklington is to the south-west of Amble.

e The railway line runs to the east of the A1068.

**3**

a Calculate the straight-line distance between Alnmouth railway station and the roundabout in nearby Hipsburn. Give your answer in metres and show your workings.

b Calculate the straight-line distance between Alnmouth railway station and the church with a spire in Warkworth. Give your answer in kilometres and show your workings.

c Calculate the distance along the B6345 between the road junction in Togston (245015) and its junction with the A1068 in Amble. Give your answer in kilometres and show your workings.

**4** Locate grid squares 2402 and 2303. Look closely at the contours in these squares and the surrounding ones and suggest why this area might be correctly described as a 'plateau'.

**5** Describe the relief in grid square 2309.

**6** Describe the course of the River Coquet and its valley from the point where it joins the map extract in grid square 2203 to its mouth just north of Amble.

**7** Look carefully at the map extract and try to find an example of each of the three main types of settlement pattern:

a nucleated

b linear

c dispersed.

For each type, give the name of the settlement and its location, using grid references. Draw simple sketch maps to support your choice. Don't forget to include scales and north points.

**8** Describe the different forms of communication between Amble and Alnmouth.

**9** What is the evidence that the area around Warkworth and Amble is popular for tourism?

**10** Use the key on page 202 to describe the stretch of coastline in the map extract.

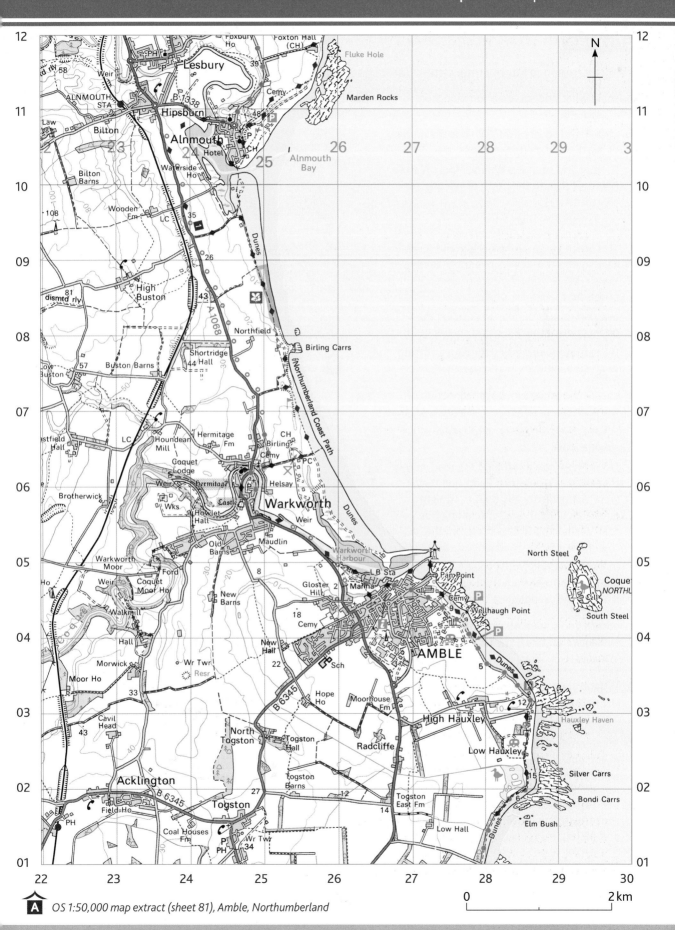

## Activities

1:25,000 Ordnance Survey map extract: Winchester (see page 203 for key to symbols)

**1** a In what grid square is Junction 9 on the M3?

b Locate grid square 4627. What is the name of the disused military camp?

c In what grid square is Winnal Moors Nature Reserve? (It is to the north of Winchester city centre.)

d What is found at 478299?

**2** a Locate St Catherine's Hill in the south-east quadrant of the map extract. What is the six-figure grid reference of the summit of this hill?

b Use compass directions to describe the orientation of the M3 in grid square 4826.

c Is the motorway passing through a cutting or an embankment in grid square 4826?

**3** Locate the roundabout at the junction of the A3090 and B3335 in grid square 4726.

a What is the six-figure grid reference of the roundabout?

b Notice how straight the B3335 road is. This is because it is a Roman Road, and they were characteristically built in straight lines. Calculate the distance along this road in a northerly direction from the roundabout to the road junction at 479295.

c In what direction would you be travelling along the A3090 from the roundabout in 4726 to the roundabout in 4627?

**4** Locate grid square 4829. What is the evidence that this is the location of Winchester's CBD?

**5** What is the evidence that the area around the railway station is part of Winchester's inner city?

**6** Stanmore in grid square 4628 is a suburban development. Do you agree? Give evidence from the map to support your view.

**7** Identify land uses typically associated with the rural–urban fringe in grid squares 4828 and 4928.

**8** Describe the drainage in the area shown in the map extract.

**9** Locate the Pilgrims' Trail in the south-east corner of the map. Draw a cross-section to show the relief of the footpath from 490276 (NW) to 500260 (SE). The first part of the cross-section has been drawn in Figure **B**. Use this to help you complete your own graph.

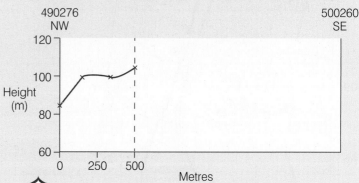

**B** *Cross-section along the Pilgrims' Trail*

**10** Several basic landscape features can be identified on Twyford Downs in the south-east corner of the map extract. Five are indicated on Figure **C**, shown by letters A–E. The five features are listed 1–5. Can you match the numbers to the letters?

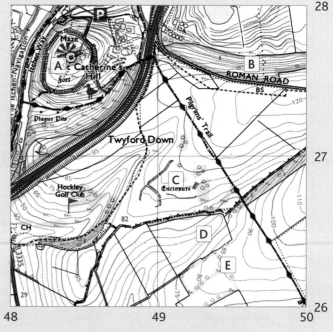

1 Ridge
2 Steep valley (a 'dry' valley typical of chalk)
3 Rounded hill
4 Spur
5 Steep slope, known as a 'scarp slope'

**C** *Extract from map D*

0                                                          1km

# ■ Key to 1:50,000 Ordnance Survey mapping (page 199)

## ROADS AND PATHS   Not necessarily rights of way

| | |
|---|---|
| Unfenced / Dual carriageway / A 689 | Primary Route |
| Footbridge / A 66 | Main road |
| B 6276 | Secondary road |
| Bridge | Road generally more than 4m wide |
| | Road generally less than 4m wide |
| | Other road, drive or track |
| | Path |
| ≫→ | Gradient : 1 in 5 and steeper  1 in 7 to 1 in 5 |

## PUBLIC RIGHTS OF WAY

| | |
|---|---|
| ·············· | Footpath |
| – – – – – | Road used as a public path |
| –·–·–·– | Bridleway |
| –+–+–+–+– | Byway open to all traffic |

## OTHER PUBLIC ACCESS

| | |
|---|---|
| • • • | Other route with public access |
| ◆ | National Trail, Long Distance Route, selected Recreational Paths |
| ○ ○ ○ | Traffic-free cycle route |
| • • • | On-road cycle route |
| 8 | National Cycle Network number |

The representation on this map of any other road, track or path is no evidence of the existence of a right of way.

## RAILWAYS

| | |
|---|---|
| | Track multiple or single |
| | Track under construction |
| | Light rapid transit system, narrow gauge or tramway |
| | Bridges, Footbridge |
| | Tunnel |
| a | Station, (a) principal |
| LC | Level crossing |
| | Viaduct |

## TOURIST INFORMATION

| | | | |
|---|---|---|---|
| Ⅹ | Camp Site | PC | Public Convenience (in rural areas) |
| | Caravan Site | | Selected places of tourist interest |
| | Garden | ℭ ℭ | Telephone, public/ motoring organisation |
| | Golf Course or links | | Viewpoint |
| i i | Information Centre, all year / seasonal | Ⅴ | Visitor centre |
| | Nature reserve | ! | Walks / Trails |
| P | Parking | ▲ | Youth Hostel |
| Ⅹ | Picnic Site | | |

## LAND FEATURES

| | |
|---|---|
| ruin | Buildings |
| | Public buildings (selected) |
| | Place of Worship { with tower / with spire, minaret or dome / without such additions } |
| ○ | Chimney or tower |
| | Cutting, embankment |
| | Quarry |
| | Spoil heap, refuse tip or dump |
| | Coniferous wood |
| | Non-coniferous wood |
| | Mixed wood |

## BOUNDARIES

| | |
|---|---|
| –+–+– | National |
| –·+·– | District |
| | County, Unitary Authority, Metropolitan District or London Borough |

## ARCHAEOLOGICAL & HISTORICAL INFORMATION

| | | | |
|---|---|---|---|
| + | Site of monument | ☆ ···· | Visible earthwork |
| · ○ | Stone monument | VILLA | Roman |
| ⚔ | Battlefield (with date) | Castle | Non-Roman |

Information provided by English Heritage for England and the Royal Commissions on the Ancient and Historical Monuments for Scotland and Wales

## WATER FEATURES

Marsh or salting · Towpath · Lock · Canal · Ford · Beacon · Weir · Footbridge · Normal tidal limit · Lake · Slopes · Cliff · Flat rock · Lighthouse (disused) · Sand · Dunes · Mud · Shingle · Lighthouse (in use) · Low water mark · High water mark

## HEIGHTS

| | |
|---|---|
| ═50═ | Contours are at 10 metres vertical interval |
| ·144 | Heights are to the nearest metre above mean sea level |

## ABBREVIATIONS

| | |
|---|---|
| CH | Clubhouse |
| MS | Milestone |
| PC | Public convenience (in rural areas) |
| TH | Town Hall, Guildhall or equivalent |
| CG | Coastguard |
| P | Post office |
| MP | Milepost |
| PH | Public house |

# ◼ Key to 1:25,000 Ordnance Survey mapping (page 201)

## ROADS AND PATHS   Not necessarily rights of way

| | |
|---|---|
| M I or A 6(M) | Motorway  Ⓢ Service area  ⑦ Junction number |
| A 35 | Dual carriageway |
| A 30 | Main road |
| B 3074 | Secondary road |
| | Narrow road with passing places |
| | Road under construction |
| | Road generally more than 4 m wide |
| | Road generally less than 4 m wide |
| | Other road, drive or track, fenced and unfenced |
| ≫ → | Gradient: steeper than 20% (1 in 5); 14% (1 in 7) to 20% (1 in 5) |
| Ferry | Ferry; Ferry P – passenger only |
| | Path |

## RAILWAYS

- Multiple track / Single track    Standard gauge
- Narrow gauge
- Light rapid transit system (LRTS), station
- Road over; road under; level crossing
- Cutting; tunnel; embankment
- Station, open to passengers; siding

## PUBLIC RIGHTS OF WAY

- ------------ Footpath
- ———————— Bridleway
- ++++++++ Byway open to all traffic
- Road used as a public path

## OTHER PUBLIC ACCESS

- •  •  • Other routes with public access (not normally shown in urban areas)
- ◆    ◆ National Trail/Long Distance Route; Recreational Route
- ------------ Permitted footpath*
- — — — Permitted bridleway*

*Footpaths and bridleways along which landowners have permitted public use but are not public rights of way. The agreement may be withdrawn

- •  •  • Traffic-free cycle route
- ☐1 National cycle network route number – traffic-free
- ◼1 National cycle network route number – on road

## TOURIST INFORMATION

| | |
|---|---|
| P P&R | Parking / Park & Ride |
| ☎ | Telephone, public / motoring organisation |
| ⛺ 🚐 | Camp site / caravan site |
| ℹ ℹ | Information centre, all year / seasonal |
| V | Visitor centre |
| 🌲 | Forestry Commission Visitor Centre |
| 🏃 | Recreation / leisure / sports centre |
| PC | Public convenience |
| ✗ | Picnic site |
| ⚑ | Golf course or links |
| 🔆 | Viewpoint |
| 🚲 | Cycle trail |
| 👪 | Country park |
| 🏛 | Museum |
| 🦋 | Nature reserve |
| 🏰 | English Heritage |
| 🦋 | National Trust |
| ☆ | Other tourist feature |

## BOUNDARIES

Administrative boundaries as notified to October 1997

- — + — + — National
- — · — · — County
- — — — Unitary Authority (UA), Metropolitan District (Met Dist), London Borough (LB) or District
- · · · · · · Civil Parish (CP) or Community (C)
- — — — Constituency (Const) or Electoral Region (ER)
- ▬▬ ▬▬ National Park boundary

## VEGETATION

Limits of vegetation are defined by positioning of symbols

- Coniferous trees
- Non-coniferous trees
- Coppice
- Orchard
- Scrub
- Bracken, heath or rough grassland
- Marsh, reeds or saltings

## GENERAL FEATURES

- ♦ Current or former with tower
- ♦ place of worship with spire, minaret or dome
- + Place of worship
- ☐ ☐ Building; Important building
- ▨ Glasshouse
- ▲ Youth hostel
- ◼ Bunkhouse/camping barn/other hostel (*selected areas only*)
- ➤ Bus or coach station
- ⛣ ⛣ ⅄ Lighthouse; disused lighthouse; beacon
- △ Ⅰ Triangulation pillar; mast
- ⋇ Windmill, with or without sails
- Ⅰ Ⅰ Wind pump; wind generator
- pylon pole Electricity transmission line
- ⟋⟋⟋ Slopes
- Gravel pit        Sand pit
- Other pit or quarry     Landfill site or slag heap

## HEIGHTS AND NATURAL FEATURES

Surface heights are to the nearest metre above mean sea level. Where two heights are shown, the first height is to the base of the triangulation pillar and the second (in brackets) to the highest natural point of the hill.

52 • Ground survey height
284 • Air survey height

Vertical face/cliff

Loose rocks   Boulders   Outcrop   Scree

Contours are at 5 metres vertical intervals

| | |
|---|---|
| ☐ | Water |
| ▨ | Mud |
| ▨ | Sand; sand & shingle |

## 12.1 Introduction

Geographers have always been concerned with investigating issues, problems and questions that arise from the interactions of people with their environments. Geographical skills are used in making decisions that are important to everyday life, such as:

- where to buy or rent a home
- where to get a job
- how to get to work or to a friend's house
- where to shop, go on holiday, or go to school.

All of these decisions involve the ability to acquire, arrange and use geographical information. Decision making is the process of evaluating alternatives and choosing a course of action to resolve a problem. It involves moving through a series of steps to consider a number of options before making the critical decision.

Typically, geographical problems may involve aspects of:

- human geography, such as the location of a new supermarket, wind farm, tourist development, housing area or factory, strategies for urban regeneration, the building of a traffic by-pass, improvements to quality of life in squatter settlements and the advantages and disadvantages of migration.
- physical geography, such as problems of coastal erosion and possible solutions, responses to an earthquake, management of river catchment areas, and the impact of deforestation and possible remedies.

### ■ Local geographical issues

You will be aware of local issues in your area. Figure **A**, for example, was taken in St Dennis in Cornwall and reflects an issue felt strongly enough by a local individual that considerable trouble has been taken to erect this sign and make a demonstration. Thorough research would be needed to find out

**A** *A local demonstration in St Dennis, Cornwall*

about this issue, to discover whether it is still a live issue or whether a decision has been made in relation to it and to explore the other side of the argument.

There is almost always another side to the argument. Some will not agree with it, but there always is one. Some arguments are NIMBY (Not In My Back Yard) type arguments. This means that those making the argument agree that the development is required, but they don't want it located next to them. Sometimes arguments are made claiming to represent a particular viewpoint, while having other underlying motives. An industrialist may claim that the proposed development is designed to help local people, but the main motive may be profit.

Even the language used can be interesting: while one side may describe a development as a 'Waste to Energy Plant', which has a positive ring to it, opponents may term it an 'Incinerator', which has much more negative connotations. A 'landmark icon' for someone in favour of a development may be an 'eyesore' for someone opposing it. Taking into consideration the various reasons behind the arguments and the emotive language used, a decision on which side has the stronger case can only be made when you have a good grasp of the issue and the competing arguments around it.

## Types of question

The decision-making exercise (DME) could be posed as a question, for example:

- How do we address conflicts in National Park management?
- Do we build a second runway for an airport?
- How should we implement a redevelopment project in a city?
- How do we choose the best route for an irrigation canal?
- Should we give permission for a wind farm on a scenic piece of land?
- How do we save bay wetlands while still allowing a city to grow?
- What are the arguments for and against a new waste incinerator?

You will be presented with a varied mix of resources related to a real world contemporary geographical issue. You may have to use geographical skills to set the issue in context to examine conflicting viewpoints about the issue. You might be required to take on a role, such as that of a planner or government minister, and describe the implications of each position.

For the Certificate course, a series of compulsory questions will be set on an issue based on one or more of the options in Units 1 and 2. The option chosen will be made available two years prior to the exam. A sources booklet, which may include maps at different scales, data sets, graphs, articles, photographs, etc. will be issued in February of the year of the exam so that you can become familiar with the topic and do some additional research. The paper will consist of a series of questions leading to a more extended piece of writing, which involves a decision with some justification.

## How to answer a decision-making question

A systematic or logical approach can be adopted when making geographical decisions:

1 Identify the problem or issue, and what needs to be achieved.
2 Assemble the evidence, organise the information and analyse the data to detect relationships and make predictions.
3 Consider the alternatives, evaluate the options and target the likely choices.
4 Make the decision based on the results of the enquiry, explain the result and implement the plan.

In justifying the decision, it is often important to consider why one option was chosen and others were rejected, which option came second and could be a reserve choice and how different options might be combined to make a better choice. Sometimes it is necessary to consider the short-term and long-term costs and benefits, whether or not the decision is an ideal solution, and whether anyone is inconvenienced by the decision and how that inconvenience might be reduced.

## Sources of geographical evidence

These could include maps on several different scales, diagrams, graphs, statistics, photos, satellite images, sketches, extracts from published materials including management plans, and quotes from different interest groups. Several sources may need to be considered before making a balanced decision. The most important thing is to take time to read and interpret the sources very carefully before answering any questions. Do not ignore any sources – all those included in the booklet should help to support or to counter a point of view or provide a context for the issue.

## Evaluating questions

In the first part of the exam you will study the sources and answer some questions on them. This will prepare you for writing the decision-making report.

When you are answering the questions it's important to read carefully what you are being asked to do. Pay attention to words and phrases like:

- explain, describe …
- compare or contrast the views …
- to what extent …
- evaluate …

You are being asked to show that you understand the information that the sources provide. To get full marks, it is important that you illustrate your answers with examples from the sources. This may include short quotations or statistics.

The final question is likely to focus on decision making. Several alternative options may be presented, each of which has advantages and disadvantages. The wording may be: 'Using all available evidence, decide which option you think is best value (or is most suitable, or which you prefer). State clearly the reasons for your choice. For one of the other options, give reasons why you rejected it.'

For example, if the DME was about coastal defences, it would be important to be fully aware of the different types of management strategies available. The advantages and disadvantages of each strategy would be assessed, then reasoned decisions made on what are felt to be the best strategies to use in the area. The choices made could be linked to the importance of the area or its value in terms of the local economy, environmental issues, housing, infrastructure and communications.

With any decision-making task the skill of being able to dispassionately evaluate an argument is important. You should not approach the exercise with a preconceived viewpoint. For example, you may generally have strong views on the environment and usually give it priority in environmental/developmental issues. Alternatively, you may hold the view that development must take precedence, even if some environmental damage is caused. However, each issue must be taken on its own merits. The viewpoints in the resources must be given their due and the strength of each position should be acknowledged. Eventually a decision will have to be made, but your justification must be sound and based on rational argument and not blind belief. You should credit strong arguments, even from a viewpoint opposed to the decision you make, but be prepared to rebut them with counterarguments. Geographical issues are complex; if they were not, they would not be contentious and not be issues. Thus, all 'sides' in an issue have a valid case to put forward and these deserve to be taken seriously and addressed with rational argument. It is possible that you can accept much of an argument provided from one viewpoint and see the strength in their case, while ultimately not agreeing that their viewpoint should prevail – the counterarguments may be stronger.

Note: there is no incorrect answer in a DME, only one which is poorly supported with evidence.

 **B** *Decision making in and around urban areas*

What is the recommended solution to problems of traffic congestion at peak times?

To what extent should new housing and shopping developments be permitted in the rural-urban fringe?

What are the most feasible strategies for dealing with the risk of floods in the urban area?

Where would be the most appropriate location for a new supermarket or industrial estate?

What are the most cost effective ways of improving leisure provision for local residents?

What are the most suitable land uses to be considered in the redevelopment of derelict sites?

What are the social, economic and environmental consequences of different routes for a traffic by–pass?

How can city planners help to promote commercial activity so as to have a minimal impact on existing architecture and layout of buildings?

 **C** *Decision making in rural environments*

Should renewable energy schemes be given planning permission in areas of natural beauty?

How can the conflicting demands of tourists, local landowners and other interest groups be resolved?

What are the most suitable options for coastal management that are affordable, effective and sustainable?

What management strategies might be pursued which conserve fragile ecosystems yet allow visitors access to honeypot sites?

What are the best ways to provide sufficient water at an acceptable cost and without harm to the environment?

Should local villages be expanded to provide new housing for a growing population?

# 12.2 Types of sources

## ■ 1 Articles and fact files

These may form the bulk of the resources provided in the booklet. There is often a section giving the background, a section giving the arguments for, another giving the arguments against and a number of news extracts and official sources. When referring to news articles and text sources, it is important to extend the explanations beyond the text itself. Very little credit is given for 'lifting' or copying directly from the source. You will be expected to have some background knowledge of the issues being discussed and to develop your ideas, using the text as a stimulus.

### Example: Features of the coastline of Holderness and Spurn Head

The Holderness coast is one of Europe's fastest eroding coastlines. Not only are settlements and agricultural land threatened, but also installations through which natural gas is brought ashore from a group of North Sea fields. Historical records show that 29 villages have fallen into the sea since Roman times. This problem continues to challenge coastal engineers, and as the pressure from population growth, economic development and recreation grows, choosing an appropriate management strategy is proving to be an increasingly difficult task.

The reason why the Holderness coast is eroding so quickly is related to the local geology. During the last ice age the north of England was covered by ice. As the ice melted it deposited huge amounts of glacial deposits. These glacial deposits actually extended the Holderness coast out into the sea.

However, the glacial deposits, known as boulder clay (or till), that make up the coast are extremely weak and vulnerable to erosion.

The clay cliffs of Holderness have limited wildlife interest. However, the sediments released from the cliffs are very important because they are carried by the sea to form the tidal mud flats of the Humber Estuary and other areas around the North Sea. To the north of Holderness, the cliffs of Flamborough Head are made of chalk. They are an important sea bird breeding site. Spurn Head on the Humber Estuary is a sand and shingle spit. It is famous for migrating wading birds and wildfowl.

### Holderness coast fact file

- Length of coastline: 61 km
- Height of cliffs: between 3–35 m
- Average annual rate of erosion: 2 m
- Material eroded per year: 3.3 million tonnes
- Geology: glacial till (72% mud, 27% sand and 1% boulders), chalk
- Physical features: chalk headland and cliffs near Flamborough, retreating clay cliffs of Holderness Bay, 6 km spit at Spurn Point.
- Factors encouraging rapid erosion:

  1 Long north-easterly fetch (over 600 km).
  2 Powerful waves when winds blow from NE.
  3 Weak unconsolidated rocks (sands and clays).
  4 Heavy rain creates saturated cliffs liable to slumping.

### Activities

1 Explain the loss of 29 villages along the Holderness coast over the past 2,000 years.

2 Explain why the release of sediments from the Holderness cliffs is significant.

3 Why is the selection of a suitable coastal management strategy for the Holderness coast 'an increasingly difficult task'?

## ■ 2 Photos

These are often included in decision-making exercises and can be useful to help form a clearer picture of the area being studied. There may sometimes be photos of models of a proposed development or artist impressions showing what the development may look like if it were to go ahead.

### Example: Coastal erosion along the Holderness coast

The images in Figure **A** illustrate some impacts of coastal erosion along the Holderness coast. Comments could relate to the sign, the nature of the cliffs, the effects on property, communications, tourism and the need for sea defences.

*Impacts of coastal erosion along the Holderness coast*

**Activities**

Study the photos in Figure **A**.

4   Give two pieces of evidence that the cliff has recently retreated.

5   Outline possible economic effects of a rapidly eroding coastline.

# 3 Annotated sketches and diagrams

These will often be included to illustrate, amongst other things, physical processes, technical details of proposed developments and methods of hard and soft engineering.

**Example: Cliff processes affecting the Holderness coast**

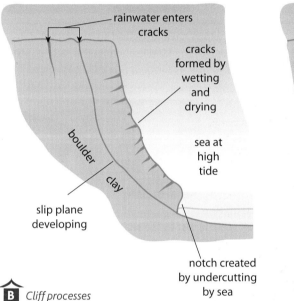
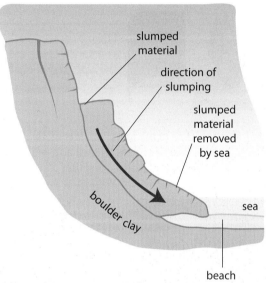

**B** *Cliff processes*

## Activities

Study Figure **B**.

**6** Explain the processes of mass movement which affect the cliffs of the Holderness coast.

**7** Why are some parts of the Holderness coastline less affected by mass movement and erosion than others?

# 4 Quotations

There may be a number of quotations summarising the views of the main groups of people involved in the proposed development. These will often give an insight into conflicting ideas about possible schemes and the potential areas of controversy. As a decision maker, you have to weigh up these ideas, and make an assessment of their validity, before making an informed choice. You could treat them as another background resource indicating the strength of feeling on both sides. Some may be neutral and balanced, but most often they represent the views of one side or the other. There is a great temptation to use these verbatim, i.e. to just write out the whole thing. This is not good practice. It may be preferable to use a particularly effective word or phrase from a quotation to support an argument being made.

## Activities

Study the quotations on page 210.

**8** The council finance director says coastal protection schemes must be 'economically worthwhile'. Explain what this means.

**9** The scientist suggests it is more important to protect the coasts bordering the southern North Sea than the coast of Holderness. Explain this view.

**10** Explain why some environmentalists might have strong opinions against the building of hard engineering sea defences.

## Example: Points of view about coastal defences

The council has to finance all coastal defence work in this area. and it is difficult to justify a major investment in coastal engineering. Some housing may be lost if the cost of protection is too high. Protection schemes must be technically sound, environmentally acceptable and economically worthwhile, with benefits greater than costs. *(Council finance director)*

Flamborough Head and Spurn Point are fantastic areas for wildlife. Any hard engineering defences could be detrimental to these wonderful environments and so any decisions made must minimise the impacts and preserve the local wildlife. *(Conservationist)*

I own a caravan park on the cliffs which have retreated so much that if no coastal defences are installed, much of my land is in danger of falling into the sea. This would be disastrous for my livelihood. *(Caravan park owner)*

I have recently moved the site of my factory to the reclaimed area on the River Humber. I pay lots of tax to the council and will expect them to protect my business in their future plans. *(Chemical plant owner)*

My farm on the edge of the cliff is being lost to the sea and I will lose everything because insurance companies refuse to insure the land and property. The erosion has been significantly worse since coastal defences were built further up the coastline at Mappleton. *(Farmer)*

The best overall solution is to avoid building in the worst affected areas, surrender land to the sea and compensate any landowners affected. Defending Holderness is not the most important thing. It is more important to ensure the long-term stability of the coasts of EU countries that border the southern North Sea. Major cities like Amsterdam and Rotterdam are located there. *(Scientist)*

Coastal protection will not work unless you build groynes. A revetment or sea-wall will give short-term protection, but without groynes no system will succeed. *(Sea-wall engineer)*

I think it is monstrous that people should lose their homes and businesses, for which they have worked and saved hard for so many years. It will mean we have nothing to leave our children. It seems to us that if the government can't get our assets, the sea will. *(Local resident)*

**C** *Views about protecting the Holderness coast*

# 5 Graphs

Graphs are sometimes used to illustrate complex data relating to the issue and are likely to be important sources of information when making decisions. The graphs may be linked to maps as indicated in Figure **D**.

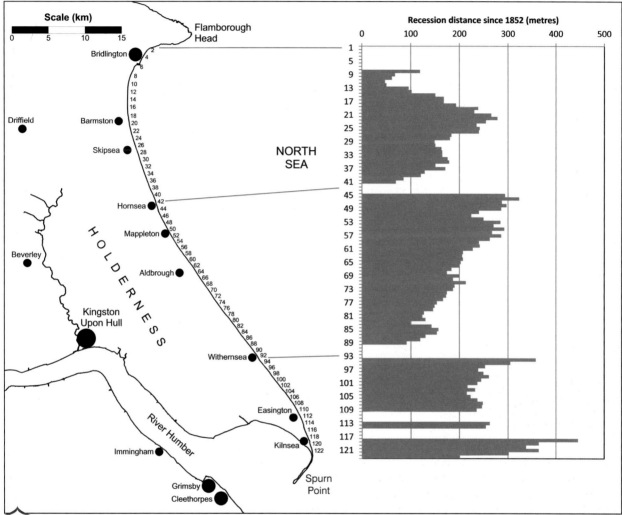

**D**   *Rates of coastal erosion along the Holderness coast. The numbers along the coast and on the left side of the graph refer to coastal monitoring profile positions established by East Riding of Yorkshire Council*

## Activities

Study Figure **D**.

**11**   What is the maximum annual rate of erosion along the Holderness coast?

**12**   Describe the general trend from north to south in the rate of coastal erosion. Suggest a reason for this trend.

**13**   Suggest why a higher rate of erosion occurs immediately to the south, and a lesser rate to the north, of the larger settlements.

## 6 Maps

There are usually small-scale maps showing general background or location of the development, and large-scale maps illustrating the scheme in more detail (Figure **E**). There may be a separate Ordnance Survey map, generally 1:50,000, but sometimes 1:25,000.

### Example: Holderness coast

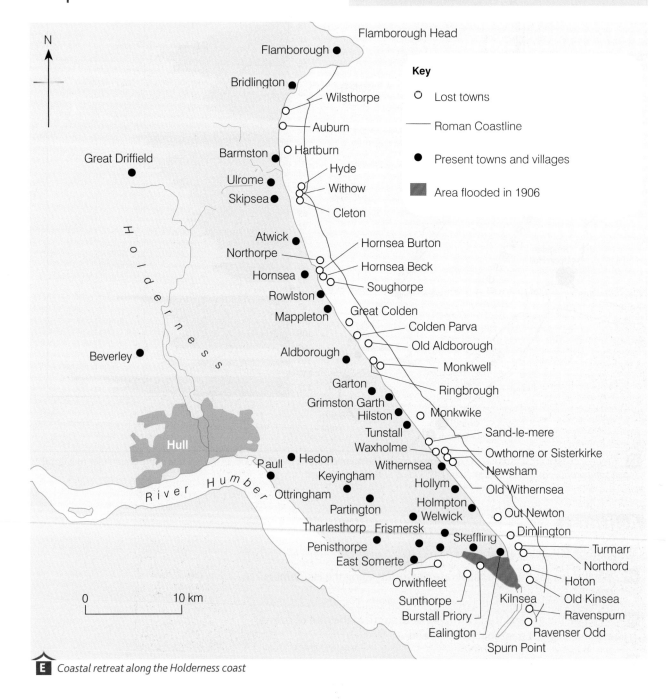

**Key**

○ Lost towns

— Roman Coastline

● Present towns and villages

▨ Area flooded in 1906

**E** *Coastal retreat along the Holderness coast*

Maps may be used in conjunction with photos (Figure **F**).

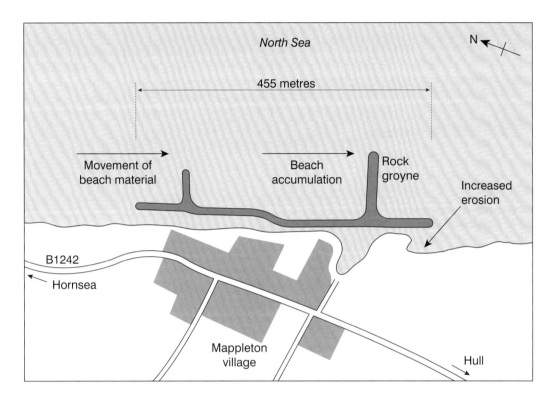

**F** *Coastal defences at Mappleton*

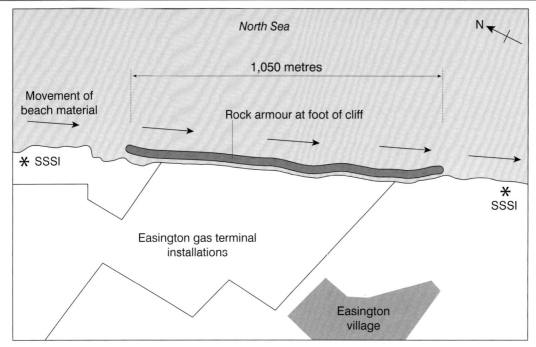

North Sea

N

1,050 metres

Movement of
beach material

Rock armour at foot of cliff

* SSSI

* SSSI

Easington gas terminal
installations

Easington
village

SSSI: Site of Special Scientific Interest

G Coastal defences at Easington

## Activities

Study Figures **F** and **G**.

16   In which compass directions was the camera facing in each of the two photos?

17   How do coastal defences at Mappleton help to protect the village? Outline one problem caused by the building of coastal defences here.

18   Suggest which factors were taken into account when building the coastal defences at Easington?

# 12.3 The main decision making exercise

## ■ Introduction

This section, usually at the end of the exam, is worth the most marks and should therefore take the most time. The question will vary according to the issue being addressed but is likely to require you to make a choice between alternatives and to justify the choice on the basis of greater overall benefits. To some extent it doesn't matter which choice is made as long as the justification is clear, balanced and detailed.

It may be necessary to consider the impacts of the proposed development. These might be economic, social or environmental, and you will probably be required to look at the benefits and drawbacks of the consequences highlighted. Economic impacts might include the employment created, the possible increase in earnings and the effects on the local economy. There may also be a loss of jobs elsewhere or a cost in other forms of economic activity. For example, a quarry will bring more jobs to an area, but might cause a reduction in jobs in tourism if it is unsightly or polluting. Social impacts are those which affect the people of the area, such as the provision of more facilities. The proposed development may have positive effects on the infrastructure (roads, power supplies, etc.), or improve the quality of life and encourage people to remain in the area. The development may also have implications for the natural environment. There may be visual, water and/or air pollution, or there may be threats to habitats of native plants and animals. Many proposed schemes will claim that they are protecting or even improving the environment.

### Example: Erosion at Holderness, on the Yorkshire coast

#### The issue

- All along the Holderness coast the cliffs are eroding rapidly.
- Some villages, such as Barmston and Skipsea with buildings on the cliff-top are in danger of falling into the sea.
- Some of the larger settlements have ageing sea defences, built largely in the early 20th century.
- What, if anything, should be done to protect the coast of Holderness?

#### Current sea defences:

A total of 11 km of coastline currently has hard coastal defences:

- Bridlington: protection here extends to 4.6 km of high masonry and concrete sea walls with groynes to stabilise the beaches. Built in the late 19th century and regularly repaired.
- Hornsea: protection for the town and its resort functions is provided by 1.86 km of concrete sea walls, groynes and rock armour. Built in the early 1900s and repaired in 1980s.
- Withernsea: there are now 2.26 km of concrete sea walls, timber groynes, rock armouring and a small offshore rock armour reef. Built in phases during the 20th century.

- Mappleton: 61,000 tonnes of rocks were used to build two groynes, a sloping revetment and graded cliffs. Built in 1991 at a cost of £1.9m.
- Easington (gas terminal and village): a 1 km long revetment was built at the base of the cliff at the terminal site, using 133,000 tonnes of rock.

## Possible sea defences

Study section 4.8 for details of coastal management, including hard and soft engineering approaches and managed retreat or coastal realignment.

## Hints for tackling this exercise

In making and justifying the choice of coastal management strategy, several criteria need to be considered. These might include the following:

- Effectiveness in tackling erosion.
- Cost. Some hard engineering schemes can cost upwards of £7,000 per metre.
- Durability, i.e. how long the proposed scheme is likely to last.
- Ease of access to the beach, which may have impacts on local residents and tourists.
- Nature/importance of the land use(s) protected. Some land uses are economically more valuable than others. Some areas have a much higher population density than others.
- Visual impact. Some hard engineering strategies are ugly. This may have implications for tourist areas in particular.
- Possible knock-on effects elsewhere. Some coastal defences have negative impacts further along the coast.
- Likelihood of getting government permission. This may depend on the strategic or economic importance of the area.

A number of well-developed points would need to be made in a thoroughly argued, balanced answer in order to achieve high marks. The answer should include coverage of the rejected schemes with a clear explanation for their rejection. It should also acknowledge some problems, as well as advantages, of the chosen option.

## ■ Decision making in the exam

It should be emphasised that most of the information needed to answer the question is provided in the sources booklet and that the exam is assessing critical understanding of the proposed developments. The knack is to efficiently select all the relevant facts and to show a clear grasp of the issue, without repeating the original text. To achieve high marks you need to state clearly the main argument and back this up with suitable evidence. The points made will need to be consistently relevant and logically structured, and show an insight into the issue. Most of the resource material will need to be used and no significant points should be omitted

### Activities

There are four possible responses to the cliff erosion along the middle part of the Holderness coast between Bridlington and Withernsea.

Which of these four responses do you think would be best?

Response chosen ........................

Give reasons for your choice and explain why you have rejected the other responses.

**1** Doing nothing and allow nature to take its course.

**2** Managed retreat, which means to defend important areas inland rather than on the coast.

**3** Building hard defences, i.e. construct sea defences from 'hard' materials not natural to beach environments. These would include the use of rock armour at the base of cliffs and offshore reefs, made from concrete filled rubber tyres.

**4** Building soft defences, i.e. working with nature and using sea defences constructed from natural coast environments.

# Should planning permission be given for an onshore wind farm?

## Background

At the end of 2012, the installed capacity of wind power in the UK was 7,100 megawatts (MW), with 360 operational wind farms and 3,900 wind turbines. The UK is ranked as the world's eighth largest producer of wind power. It became the world leader of offshore wind power generation in October 2008 when it overtook Denmark. It also has the largest offshore wind farm in the world, the Thanet wind farm, located off the Kent coast.

The UK urgently needs an increase in renewable energy provision but, unfortunately, wind turbines are seen by some to be a symbol of visual pollution, disfiguring the environment. Wind turbines are grouped together on high ground as wind farms. The turbines must be placed apart so that they do not shelter each other from the wind. They must also be on raised ground or by the coast in an area that is not obstructed by trees or buildings so they can get the most wind. Most wind farms are found in open spaces in rural areas but they can also be located offshore. The planning of new wind farms is highly contentious and opinions can be sharply divided. The subject of renewable energy is often in the news as these headlines indicate:

### Wind power: new poll finds 66 per cent of UK public in favour

Majority agree that onshore wind energy plays a role in a balanced UK electricity mix alongside gas, nuclear, cleaner coal and other forms of renewable energy.

### UK on course to hit 2020 green energy targets

During the period from July 2011 to June 2012, renewable power output grew by 27 per cent, while overall renewable energy accounted for over 10 per cent of total electricity. There was a 60 per cent increase in offshore wind capacity, and a five-fold increase in solar photovoltaic capacity.

### Wind energy claims are just a lot of hot air

Over the next eight years, the UK needs to spend £100 billion on building 30,000 useless, unreliable and heavily subsidised wind turbines, according to an anti-wind farm pressure group.

### Nine out of 10 people want more renewable energy

Just 2 per cent backed more gas power in the latest UK opinion poll to show overwhelming support for clean energy.

## The issue

Cornwall has some of the highest wind speeds of any European country – on average 6.5 metres per second. The first commercial wind farm in the UK was commissioned at Delabole in Cornwall in 1991 with an installed capacity of 4 MW. Since then, nine wind farms have become operational in Cornwall with a capacity of 58 MW and planning approval is being sought for several further schemes.

One of these is situated at Truthan Barton, near Truro.

While the need for green energy is great, this development would, in the opinion of many, have too great an impact on the surrounding land and villages. Should planning permission be given for Truthan Barton?

### Resource 1: Renewable energy targets

The Renewable Energy Directive sets a target for the UK to achieve 15 per cent of its energy consumption from renewable sources by 2020. The government has also asked the Committee on Climate Change (CCC) for advice on the level of ambition for renewables in 2020 and beyond, taking into account cost, technical potential, environmental impact and practical delivery. The government's Renewable Energy Strategy lead scenario suggests that by 2020 about 30 per cent or more of our electricity could come from renewable sources, compared with 10 per cent today.

The UK has more than 4 GW of installed onshore wind capacity in operation (generating approximately 7 TWh of electricity annually). Onshore wind could contribute up to around 13 GW by 2020. Achieving this level of capacity equates to an annual growth rate of 13 per cent.

The UK is the global leader for offshore wind energy with 1.3 GW of operational capacity across 15 wind farms (which generated over 3 TWh during 2010). The UK is well placed to continue this lead role to 2020 and beyond. Up to 18 GW could be deployed by 2020.

### Resource 2: Wind Farms in the UK, 2012

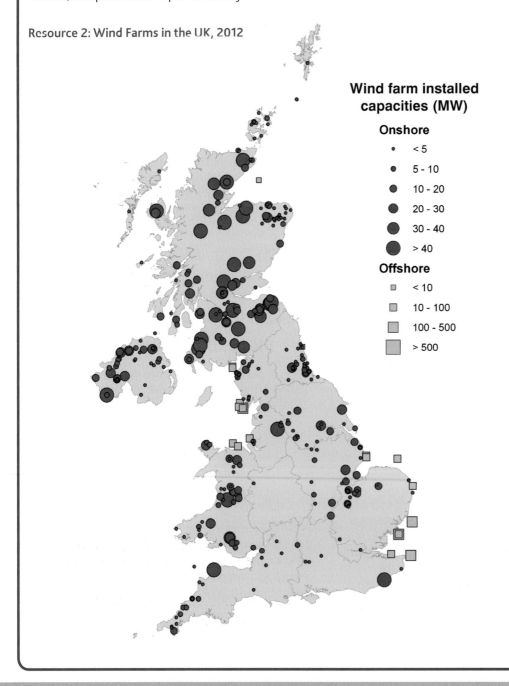

**Wind farm installed capacities (MW)**

**Onshore**
- • < 5
- • 5 - 10
- ● 10 - 20
- ● 20 - 30
- ● 30 - 40
- ● > 40

**Offshore**
- ▫ < 10
- ◻ 10 - 100
- ◻ 100 - 500
- ◻ > 500

## Activities

**5** Using Resource **1**, suggest reasons why the government considers that an increasing amount of our energy should come from renewable sources?

*Hint: think about the fact that fossil fuels are running out, are expensive to mine, cause pollution, adding to greenhouse gases, meeting international commitments and the benefits of renewable energy.*

**6** a Using Resource **2**, describe the distribution of wind farms in the UK, both onshore and offshore.

*Hint: use specific places and regions, so an atlas will be needed. Look at trends where offshore farms are compared to onshore farms.*

b Explain why some parts of the country have virtually no wind farms.

*Hint: planning law, physical factors, type of land use.*

**7** Give two reasons why the UK will face an energy gap in the next 10 years.

*Hint: increased demand and what is happening to fossil fuels and nuclear power stations.*

**Resource 3: The relative costs of power generation**

| Energy source | Nuclear | Coal | Onshore wind | Offshore wind |
|---|---|---|---|---|
| Cost (pence) per Kwh | 2.5 | 2.3 | 3.7–5.4 | 5.5–7.2 |

**Resource 4: Potential amounts of alternative energy in the UK using present technology**

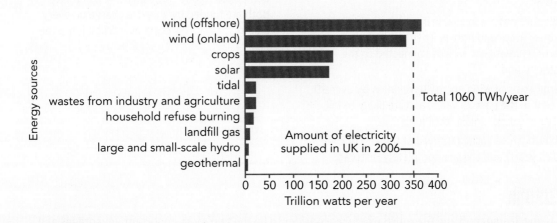

## Activities

**8** Describe the difference between 'onshore' and 'offshore' wind farms.

**9** Explain why the cost of offshore wind power generation is greater than that of onshore wind power.

**10** Why do you think that wind may have an important role to play in meeting UK energy needs in the future?

*Hint: think of the location of the UK, climate change and decreasing costs.*

Case study

## Resource 5: Advantages of wind energy

- Wind power is low cost after the initial production and installation.
- Wind power is clean. There is no pollution or carbon dioxide after the initial setup, apart from the comparatively minor emissions produced to manufacture, transport, erect and maintain the turbines.
- Wind is a renewable and sustainable resource.
- Wind power will become cheaper than fossil fuel in the next few years when the price of carbon is added to coal and oil.
- Wind turbines are self-sufficient, as they are low maintenance, with few moving parts and are easy to repair.
- The land beneath the turbines can still be used for farming or other purposes.
- Wind energy reduces dependence on imported oil and gas.
- The UK needs to meet a target to generate 15 per cent of energy from renewables by 2020 as part of global plans to fight climate change.
- Manufacturing and installing wind farms in the UK provides 'green jobs'.
- Many people find wind farms an interesting feature of the landscape. Some have visitor centres for tourists.
- Single turbines are available in a variety of sizes to suit businesses and homes. They can be installed in remote areas that are not connected to the electricity grid.
- Overall costs of wind energy have fallen by over 50 per cent in the past 25 years, and are likely to fall further in future, unlike fossil fuels.
- Turbines can be easily removed at the end of their life (25 years) and the site completely restored.

## Resource 6: Disadvantages of wind energy

- The strength of the wind is not constant and it varies from zero to storm force.
- Electricity from wind farms is unreliable and cannot be stored. Wind is an intermittent and unpredictable energy source. Wind farms may require at least 90 per cent backup from reliable controllable energy sources.
- Some people see large wind turbines as unsightly structures and not pleasant or interesting to look at. They believe that wind turbines degrade the landscape.
- Wind turbines generate noise. Some people who live near wind turbines find this noise annoying. Some of those people get stressed by it.
- A large number of turbines are needed to produce a moderate amount of energy.
- Tourism may be affected if there are too many wind farms in areas of outstanding natural beauty.
- Wind farms tend to have poor access as the windiest places are in remote rural areas. This makes it hard to maintain them, especially in the winter.
- Wind farms may reduce landscape value, kill birds and compromise wildlife habitats.
- No developer will build wind farms without subsidies. In the UK, electricity companies are compelled to buy this expensive electricity. Both are funded by an extra charge on every electricity bill. This costs over £1 billion a year.

### Activity

11  Using Resources **5** and **6**, explain what you consider to be the two most important advantages and disadvantages of wind power. Justify your choices.

*Hint: each point needs to be developed by giving two linked reasons for each advantage and disadvantage.*

**Resource 7: Proposed site for a wind farm at Truthan Barton, Cornwall**

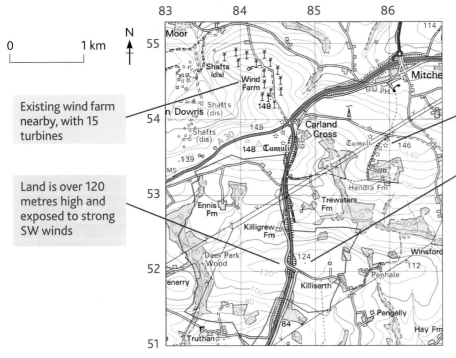

Existing wind farm nearby, with 15 turbines

Site is close to existing power lines

Land is over 120 metres high and exposed to strong SW winds

Proposed site for Truthan Barton wind farm. Seven turbines planned, with a maximum potential generating capacity of 17.5 MW. Each turbine would have a maximum height of 120 m. Capacity to generate electricity for about 10,000 homes.

*OS 1:50,000 map extract (sheet 200)*

**Resource 8: View looking south west towards proposed site at Truthan Barton**

**Activity**

**12** Using Resources **7** and **8** describe the features of the Truthan Barton site for the building of a wind farm.

*Hint: comment on the relief of the land, height, proximity to settlements and communications.*

Case study

Resource 9: Points of view about wind farming

It is difficult to think of a worse act of cultural vandalism than putting turbines near the villages of Trispen and St Erme, and the surrounding medieval landscape. The 100-metre towers would be a terrible eyesore and property prices will plummet. *(Local councillor)*

We must use more renewable energy. The UK depends too much on fossil fuels, which are running out and becoming increasingly expensive. Burning them contributes to global warming. Wind farms produce a clean form of electricity and will make a real contribution to our supplies. *(An environmental group)*

It would be fine for my land to be used for the wind farm. As the wind turbines are very tall and take up little space, I can still keep my cows here but also earn a decent subsidy from the wind farm company. *(Local farmer)*

Wind energy is a popular form of energy. A recent survey showed that 90 per cent of tourists to Cornwall didn't mind seeing wind farms in the county and in fact 7 out of 10 people said they'd be interested in visiting a wind farm information centre as part of their holiday. *(A sustainable energy company)*

These turbines would be so much bigger than anything we have got in Cornwall – the scale is quite breathtaking and no one would want to live close to one. Profit-hungry renewable energy developers are simply looking to cash in on taxpayer funded subsidies and credits. Wind turbines are unreliable, unsightly, inefficient and costly. Enough is enough. *(Chairman of local pressure group)*

The Government recognises the many benefits of renewable energy and is actively supporting the development of onshore and offshore wind farms to help tackle climate change and encourage a more sustainable approach to energy use. We do believe however that local people should be fully consulted about new developments and that the full planning process should be adhered to. *(A Government official)*

The noise from the wind farm will make people's lives miserable. They will get headaches all the time and won't be able to sleep properly. The turbines are ugly to look at and don't operate half the time. *(Local resident)*

Wind turbines produce an adaptable form of energy that can provide electricity for any consumption level. The cost will continue to fall as turbine technology improves. *(University lecturer)*

Wind turbines on average only give 20 per cent of their maximum power over a year. You would need over 25,000 turbines to replace one coal-fired power station. Onshore wind is the costliest form of electricity for the least output, and the consumer will have to bear these costs. *(Energy expert)*

## Activities

**13** Choose two of the objectors to wind farms and explain why they may have a good case for not liking wind farms.

*Hint: refer to the source but develop ideas in your own words.*

**14** Choose one of the objectors and explain why their argument is not very convincing as a reason for rejecting wind farms.

**Resource 10: News article about wind farming in Cornwall**

# Wind farm war is fought out at planning appeal

A developer facing fierce opposition to his plans for a wind farm near Truro has vowed to continue his fight even if he loses a planning appeal being heard this week. He has spent the past eight years and about £100,000 trying to win approval for the renewable energy farm on land at Truthan Barton, St Erme. His scheme horrified residents, who fear that it would ruin the unspoilt landscape and they formed an opposing campaign group. Its chairman said it raised £50,000 in donations within the parish to challenge the wind farm, winning the right to speak at the inquiry.

'We represent 90 per cent of St Erme people (from a door-to-door poll) and are supporting Cornwall Council's argument that this would have an adverse impact on the landscape and heritage of Truthan Barton. We also want to stress the effect on residents' amenities, their health and wellbeing with concerns around noise, sleep and the flicker effect (caused by the rotation of wind turbine blades).'

The developer was forced to scale down his plans after a high court judgment upheld Cornwall Council's refusal of planning permission. His latest application for seven 120-metre turbines – taller than Truro Cathedral – was turned down by Cornwall Council and is now subject to a six-day planning inquiry at St Erme's community centre. Opening the hearing on Tuesday, the planning inspector said it would focus on the turbines' individual and cumulative impacts on the landscape, character and its visual appearance. It would also look at its cultural and heritage assets and the residents' amenities and whether the harm outweighed any benefits from the scheme.

The developer said: 'There is a small vocal minority who are, in my view, overreacting. Once up, no one would notice them. These people are riling everyone, spreading panic that the turbines are too noisy and will bring house prices down.' Watching the hearing, one Cornwall councillor backed the development, saying: 'We need these projects, we're facing an energy crisis. This will make a significant contribution to our energy needs.'

During opening arguments, the solicitor for the residents gave a strong defence of Truthan Barton, saying: 'It is important to the locals as a precious piece of mid-Cornish countryside that has remained unchanged since Medieval times.' A spokesperson for the Realistic Energy Forum South West accused the council of 'inconceivable ignorance' claiming wind turbines are inefficient and ruin the landscape. He said: 'I'm extremely alarmed. Who wants those industrial monsters near their homes? The council is just chasing money to the detriment of Cornwall.'

A representative of Cornwall Council said green energy should not be 'at any cost' and that full consideration would be given to the views of local people.

## The decision making task

Use any of the resources provided and your own knowledge.

Coronation Power wishes to develop a wind farm at Truthan Barton in Cornwall in order to produce more renewable energy. Many people object to this plan and it is a major source of controversy. There are four possible ways forward for the local authorities:

### Option 1

Turn down the wind farm plan completely.

### Option 2

Allow the full development of the wind farm plan as suggested by Coronation Power.

### Option 3

Allow the development of a smaller wind farm with 2–3 turbines.

### Option 4

Allow the company to build a large wind farm in an alternative location in Cornwall.

a Choose one of the above options and explain why you chose it.

b Give one reason why you rejected each of the other three options.

c Explain one possible disadvantage of your chosen option.

d Explain why any two of your rejected options may still be considered sustainable.

# 13 Assessment

## 13.1 Paper 1: Dynamic physical world

### 1 Tectonic activity and hazards

Study Figure **A** on page 17 showing the distribution of volcanoes.

**a**  
  i    Describe the world distribution of volcanoes. *(3 marks)*

  ii   Explain why there are no active volcanoes in the UK. *(1 mark)*

  iii  The volcanoes of Hawaii (for example, Mauna Loa) are hot spot volcanoes. Explain how a hot spot volcano is formed. *(2 marks)*

  iv  Mt St Helens is a composite volcano. Draw a labelled diagram to show the main features of a composite volcano. *(4 marks)*

  v   Describe the advantages for local people of living in an area of volcanic activity. *(3 marks)*

**b**    Study Figures **A**, **B** and **C** on pages 12 and 13. Using the photos and your own knowledge, explain how people adapt to challenging physical conditions in areas of tectonic activity. *(4 marks)*

**c**    Study Figure **B** on page 32, a photograph showing the impact of a tsunami in Japan. Using a case study, describe the short-term and long-term effects of a tsunami. *(8 marks)*

### 2 Ecosystems and global environments

**a**    Study the article below about deforestation in the Brazilian rainforest.

## Growing Demand for Soya Beans Threatens Amazon Rainforest

Deforestation of the Brazilian Amazon rainforest has increased almost sixfold, recent data suggests. Satellite images show deforestation increased from 103 sq km in March and April 2010 to 593 sq km (229 sq miles) in the same period of 2011. Much of the destruction has been in Mato Grosso state, the centre of soya farming in Brazil.

Deforestation brings with it many consequences-air and water pollution, soil erosion, malaria epidemics, the release of carbon dioxide into the atmosphere, the eviction and decimation of indigenous Indian tribes, the loss of biodiversity through extinction of plants and animals and the potential loss of cures for many human diseases. Fewer rainforests mean less rain, less oxygen for us to breathe, and an increased threat from global warming.

Responsibility for this deforestation lies with logging companies, cattle ranchers, and mining corporations, but the greatest emerging threat to Amazon rainforests and communities is industrial soya plantations which are claiming large tracts of land. The Brazilian government gives financial incentives to increase soya bean production, and Brazil is now the world's second largest producer of this profitable crop. The leading importers are European countries and China, with most consumed as animal feed and soya bean oil.

Brazil has discussed reducing deforestation 80 per cent by 2020 as part of its contribution to lowering global carbon emissions. Unfortunately, if soya bean consumption continues to climb, the economic pressures to clear more land could make this very difficult.

|     |     |     |                                                                                                                                                                   |             |
| --- | --- | --- | ----------------------------------------------------------------------------------------------------------------------------------------------------------------- | ----------- |
|     |     | i   | What is meant by deforestation?                                                                                                                                   | *(1 mark)*  |
|     |     | ii  | State **two** causes of deforestation other than those mentioned in the article above.                                                                            | *(2 marks)* |
|     |     | iii | Explain why a lot of soya beans are now being grown in Brazil.                                                                                                     | *(1 mark)*  |
|     |     | iv  | Using the article above and your own knowledge, explain why environmentalists say that Brazil should preserve more of its rainforest.                              | *(4 marks)* |
|     |     | v   | Outline how deforestation can lead to problems of soil erosion.                                                                                                    | *(2 marks)* |
|     | b   |     | Study Figure **A** on page 37, a map of world biomes. Describe the world distribution of hot deserts.                                                              | *(3 marks)* |
|     | c   |     | Study photos 3 and 4 in Figure **H** on page 41, showing features of desert vegetation. Using Figure **H** and your own knowledge, explain how vegetation is adapted to climate in hot desert areas. | *(4 marks)* |
|     | d   |     | Use a case study of a hot desert area to describe how people try to manage the area in a sustainable way.                                                          | *(8 marks)* |

## 3  River processes and pressures

|     |     |                                                                                                                                                  |             |
| --- | --- | ------------------------------------------------------------------------------------------------------------------------------------------------ | ----------- |
| a   | i   | Study Figure **B** on page 58, a photo of High Force, a waterfall on the River Tees. Describe the features of the waterfall.                       | *(3 marks)* |
|     | ii  | Draw a labelled diagram(s) to explain the formation of a waterfall.                                                                              | *(4 marks)* |
| b   | i   | Describe how a river transports its load.                                                                                                        | *(3 marks)* |
|     | ii  | Name **one** physical feature resulting from the deposition of a river's load.                                                                    | *(1 mark)*  |
| c   |     | For an example of a dam and reservoir scheme you have studied, describe the location and characteristics of the scheme.                           | *(4 marks)* |
| d   | i   | Outline **two** reasons why some rivers are liable to flood.                                                                                      | *(2 marks)* |
|     | ii  | Using a case study, describe the effects of a river flood.                                                                                        | *(8 marks)* |

## 4  Coastal processes and pressures

|     |     |                                                                                                                           |             |
| --- | --- | ------------------------------------------------------------------------------------------------------------------------- | ----------- |
| a   |     | Give **two** differences between destructive waves and constructive waves.                                                | *(2 marks)* |
| b   | i   | Study the photograph of Hurst Castle Spit below.                                                                          | *(3 marks)* |

|     |     |                                                                                                                           |             |
| --- | --- | ------------------------------------------------------------------------------------------------------------------------- | ----------- |
|     | ii  | Explain how the spit was formed and how it might change in the future.                                                     | *(4 marks)* |
| c   | i   | Outline why sea level is expected to rise.                                                                                 | *(2 marks)* |
|     | ii  | Describe some economic effects of coastal flooding.                                                                       | *(4 marks)* |
| d   |     | Different methods are used to protect coastal areas from the effects of physical processes.                               |             |
|     | i   | What is soft coastal engineering?                                                                                         | *(2 marks)* |
|     | ii  | For an example you have studied, explain the coastal management methods that are being used to deal with physical processes. | *(8 marks)* |

# 13.2   Paper 2: Global human issues

## 5   Contemporary population issues

**a**   i   People move from one country to another. Describe some advantages to the two countries involved. *(3 marks)*

     ii   Suggest possible push and pull factors that encourage migrant workers to leave their home countries. *(3 marks)*

**b**   i   What is the meaning of the term 'birth rate'? *(1 mark)*

     ii   Study Figure **A** on page 99, a graph showing the demographic transition model. Explain the changes that take place in the birth and death rates in Stages 2 and 3. *(4 marks)*

     iii   Give **two** ways that population structure changes when a country has entered Stage 5 of the demographic transition model. *(2 marks)*

**c**   Many MEDCs (more economically developed countries) have an ageing population. Describe how governments are attempting to manage the problem of an ageing population. *(4 marks)*

**d**   Using **one** or more examples, explain what countries are doing to cope with rapid population increase. *(8 marks)*

## 6   Contemporary issues in urban settlements

**a**   i   What is the meaning of the term 'urbanisation'? *(1 mark)*

     ii   Study Figure **A** on page 130, photos of squatter settlements in LEDCs (less economically developed countries). Using Figure **A** and your own knowledge, explain why the conditions shown in the photos are found in squatter settlements. *(4 marks)*

     iii   Give **two** locations within a poor world city where squatter settlements are often found. *(2 marks)*

     iv   Describe ways in which the lives of people living in some squatter settlements have been improved. *(4 marks)*

     v   Explain why the disposal of waste in some cities in the poor world is very difficult. *(3 marks)*

**b**   Compare the characteristics of the inner city and the CBD (central business district) in MEDC cities. *(3 marks)*

**c**   Using a named urban area, describe what has been done either to improve the CBD or to regenerate the inner city. *(8 marks)*

## 7   Globalisation in the contemporary world

**a**   i   What is a transnational corporation (TNC)? *(1 mark)*

     ii   Study Map **B** on page 148. Describe the global distribution of Toyota's factories. *(3 marks)*

     iii   Toyota is planning to build another new car-making factory in China, which might benefit the area in which it is built. Describe some of the benefits. *(4 marks)*

     iv   Give **two** possible advantages to TNCs of locating factories in LEDCs. *(2 marks)*

**b**   i   Explain the recent huge increase in the global demand for energy. *(3 marks)*

     ii   Increased energy use can cause global environmental impacts. Using the map on page 228, describe some likely environmental impacts of global warming. *(4 marks)*

     iii   For **one** type of renewable energy you have studied, describe its advantages and disadvantages. *(8 marks)*

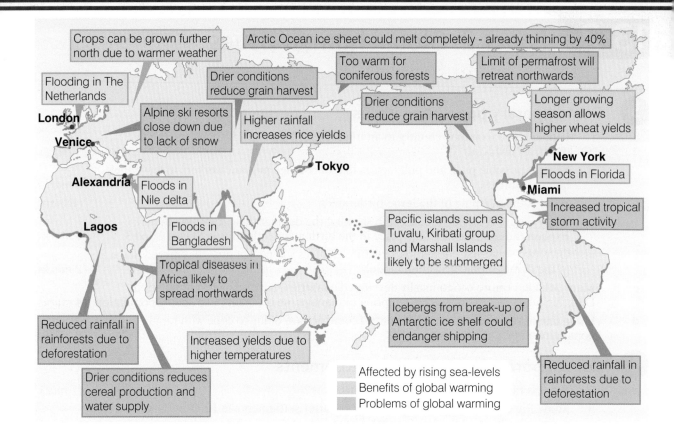

Crops can be grown further north due to warmer weather

Arctic Ocean ice sheet could melt completely - already thinning by 40%

Too warm for coniferous forests

Limit of permafrost will retreat northwards

Flooding in The Netherlands

Drier conditions reduce grain harvest

Drier conditions reduce grain harvest

Longer growing season allows higher wheat yields

London

Alpine ski resorts close down due to lack of snow

Higher rainfall increases rice yields

Venice

Tokyo

New York

Floods in Florida

Alexandria

Floods in Nile delta

Miami

Increased tropical storm activity

Lagos

Floods in Bangladesh

Pacific islands such as Tuvalu, Kiribati group and Marshall Islands likely to be submerged

Tropical diseases in Africa likely to spread northwards

Reduced rainfall in rainforests due to deforestation

Increased yields due to higher temperatures

Icebergs from break-up of Antarctic ice shelf could endanger shipping

Reduced rainfall in rainforests due to deforestation

Drier conditions reduces cereal production and water supply

Affected by rising sea-levels
Benefits of global warming
Problems of global warming

## 8 Contemporary issues in tourism

**a** **i** Study Figure **A** on page 163. Explain why some parts of the world have experienced a huge growth in tourist numbers. *(3 marks)*

**ii** In some years the number of international tourists drops. Suggest why this might happen. *(2 marks)*

**iii** Mountainous areas have been developed as tourist areas. Describe the physical and human attractions which have led to their development as tourist areas. *(4 marks)*

**b** **i** Study photos **A**, **B**, **C** and **D** on page 180. Explain what attracts people to extreme environments. *(2 marks)*

**ii** For **one** area of extreme tourism you have studied, discuss the extent to which the area can cope with the development of a tourist industry. *(4 marks)*

**c** **i** Suggest why the north coast of Jamaica shown in Figure **D** on page 178 is an example of mass tourism. *(2 marks)*

**ii** For **one** tropical area of mass tourism you have studied, describe the positive and negative impacts (effects) of tourism on the economy and environment. *(8 marks)*

## 13.3   Paper 3: Applied geographical skills and decision making

### ■ Section A

1    Study the 1:50,000 Ordnance Survey map extract below, showing part of the Dorset coast.

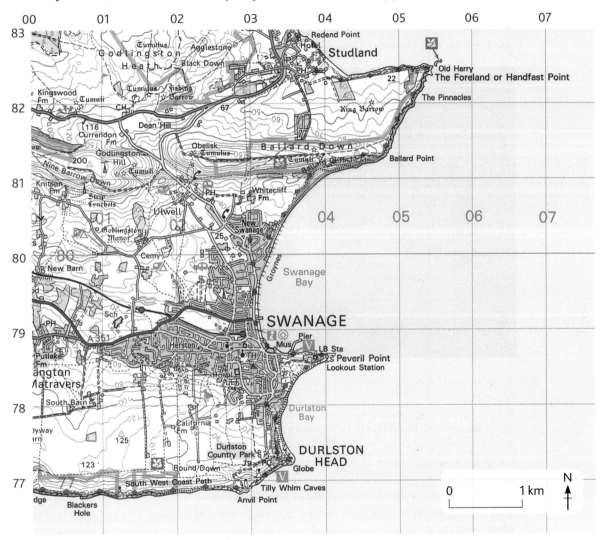

a    Match the following grid references to the correct physical feature in the table below:

005769        040827        013776

| Physical feature | Grid reference |
| --- | --- |
| Plateau | |
| Steep cliff | |
| Sandy beach | |

*(2 marks)*

**b** Locate Swanage Bay. What is the width of the bay from Ballard Point to Peveril Point? *(1 mark)*

**c** Which **two** descriptions of the caravan park in grid square 0280 are correct?

Close to a railway.
Next to a public house.
At the base of a steep south-facing slope.
Near a river. *(2 marks)*

**d** Give **three** pieces of map evidence to show that this whole area is popular with tourists. *(3 marks)*

**e** Suggest **one** advantage and **one** disadvantage of the location of Knitson Farm in grid square 0080. *(2 marks)*

**f** Contrast the relief of the land in grid square 0281 with that of 0077. *(2 marks)*

**g** A large new housing estate is planned for grid square 0277. Do you think this is a good site for new housing? Give **two** pieces of map evidence to support your answer. *(2 marks)*

**h** Study the photo of Old Harry Stacks above. Using the map extract and the photo, give the six-figure grid reference of feature **A** shown on the photo. *(1 mark)*

**i** Using the map and photo, describe features of the coastline shown in grid squares 0582 and 0482. *(4 marks)*

**2** Study the population pyramid for Swanage in Dorset on page 231.

**a** How many males are there between the ages of 25 and 29? *(1 mark)*

**b** Which age group has the largest number of people? *(1 mark)*

**c** What does the graph show about population over the age of 70? *(2 marks)*

**d** Suggest possible reasons for your answer to **c** above. *(2 marks)*

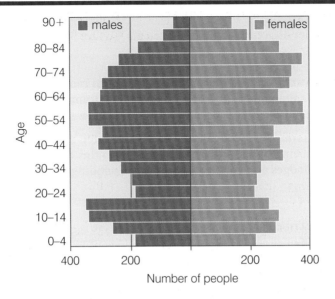

3   The following data was collected as part of a survey about tourists visiting Swanage.

| Home area of visitors to Swanage | Number of visitors | Average distance travelled (km) |
|---|---|---|
| South East England | 43 | 85 |
| South West England | 29 | 140 |
| English Midlands | 38 | 280 |
| North West England | 16 | 375 |
| North East England | 11 | 510 |
| Scotland | 6 | 780 |
| Wales | 10 | 340 |
| Outside the UK | 6 | 1200 |

a   Draw a scattergraph to show the relationship between the number of visitors and average distance travelled. *(2 marks)*

b   Describe the relationship between the number of visitors to Swanage and the distance travelled. *(1 mark)*

c   What type of map could be drawn to show the home areas of visitors and their numbers? *(1 mark)*

4   Three types of graph used in geography are a triangular graph, a bar graph and a line graph.

For each of the following three examples, (a–c), choose the most appropriate type of graph from the list above. Name the type of graph and draw a simple sketch identifying its key features.

a   A graph to show rainfall totals for months of the year.

b   A graph to show change in the population over a number of years.

c   A graph to show the percentages of primary, secondary and tertiary employment in a group of countries. *(6 marks)*

# ■ Section B

## Background to the issue

Iwokrama Forest in central Guyana is one of the last pristine tropical rainforests in the world. It is an area of extraordinary beauty with a tremendous diversity of natural environments. These include wetlands and lakes, lowland tropical forests and savannahs, and dense rainforests on steep-sided 1,000 m high mountains.

An ecotourism lodge, the Iwokrama River Lodge, has been constructed and the tourist facilities in the area have been upgraded. It is now being proposed that an extension to this ecotourism development should take place, consisting of six further lodges and communal facilities, including a new larger building.

## Questions

1   Using Resource **2**, describe the location of Iwokrama Forest.                     *(1 mark)*
2   Measure the straight-line distance from the northern edge of Iwokrama Forest to Georgetown.                                                                        *(1 mark)*
3   Using Resource **3**, describe the relief of the area shown in this photo of Iwokrama.    *(2 marks)*
4   Suggest reasons for the construction of the road shown in Resource **3**.           *(2 marks)*

Study Resources **4**, **5** and **6** that show some of the characteristics of sustainable forestry in Iwokrama.

5   Explain why it is important that the rainforest is managed in a sustainable way.     *(4 marks)*
6   What evidence is there to suggest that logging in Iwokrama is 'low impact', causing only minimal damage to the forest?                                                  *(3 marks)*

The Iwokrama International Centre for Rainforest Conservation and Development wishes to expand ecotourism at the Iwokrama River Lodge. The plan involves constructing a further six buildings, each capable of housing up to 10 people. In addition, a large communal building will be constructed to provide a meeting room, kitchen, dining room, washing and toilet facilities. The new facilities will be aimed at the luxury end of the market.

Study Resources **7** to **10**

7   a   Identify the attractions for ecotourism in the rainforest.                       *(2 marks)*
    b   Outline the potential advantages of expanding ecotourism.                        *(2 marks)*
    c   Suggest the likely environmental and social problems that may be associated with an expansion of the River Lodge.                                                    *(4 marks)*
    d   Explain how the following principles of ecotourism could be incorporated into the expanded River Lodge.
        •   Use of local building materials.
        •   Minimise the need for lighting and heating, and no use of fossil fuels.
        •   The need to conserve water and minimise waste.
        •   Use and involvement of local communities.
        •   Need to conserve the local natural environment.                            *(5 marks)*
    e   Do you think the proposed development should take place at River Lodge? Justify your answer.                                                                       *(9 marks)*

## Resources

### Resource 1: Iwokrama Forest – diversity of species, deforestation and conservation

The Iwokrama Forest is home to 475 species of birds and the highest number of species of fish (over 400) and bats (over 90) recorded in any area of a comparable size in the world. The area is also the homeland of the Makushi people who continue to live in the area and use the forest and wetland resources.

However, the forest is under serious threat from deforestation. In 1995, over 90 per cent of Guyana's land area was forested. Just a decade later, this figure had decreased to 75 per cent. This is due to an alarming increase in the rate of deforestation to create land for farming, mining and road building. Much of the wood has been used for logging or to provide fuelwood.

The Iwokrama Project is one of a number of conservation projects that aims to save the rainforest by exploiting its potential in sustainable ways and supporting the needs of the local community. Nearly one million acres (371,000 hectares) of forest has been given special protection by the Guyanese government, who want it to be used for science, learning and as a showcase for sustainability. The area is also being developed for ecotourism, which is proving particularly popular with bird-watching groups from around the world.

### Resource 2: Location map of Iwokrama Forest, Guyana

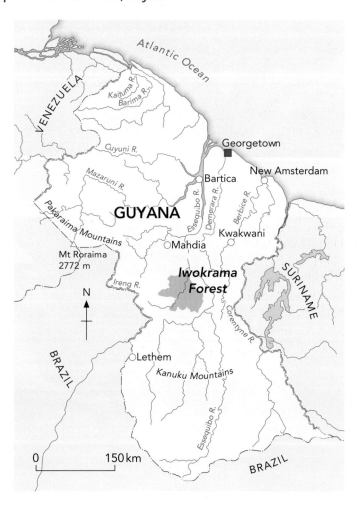

**Resource 3: Aerial photo of Iwokrama Forest, Guyana**

**Resource 4: Sustainable logging in Iwokrama Forest, Guyana**

**Resource 5: Removal of selected trees from Iwokrama Forest, Guyana**

**Resource 6: Sustainable management**

A management operation certified by the Forest Stewardship Council (FSC) has been implemented to encourage sustainable logging practices such as 'reduced impact logging'. This involves detailed mapping of the area to select individual trees for logging. Trained professionals fell the trees in a particular direction so as not to cause widespread damage. The fallen trees are then removed from the forest with minimum impact and a programme of re-planting undertaken. A strong emphasis is placed on involving local people and training them to become foresters and rangers.

**Resource 7: Iwokrama River Lodge, Guyana**

## Resource 8: Ecotourism

The Iwokrama Forest and the Rupununi Wetlands and savannahs offer visitors the opportunity for an exceptional natural and cultural experience set in a learning context.

Iwokrama is a place for all ages and all interests and you choose what you want to do. By staying at Iwokrama you are directly contributing to the communities in and surrounding the forest and to the conservation of what lies within. You will also contribute towards the development of an eco-friendly sustainable tourism model which can be shared locally, nationally and with the international community.

The Iwokrama Forest is a vast wilderness of nearly one million acres – a protected area and homeland of the Makushi people who have used the forest for generations. A unique blend of technology, ancient culture, and traditional knowledge, Iwokrama fulfils its mandate and promise to the world.

Come and retrace the footsteps of the Makushi people, wander our extensive trail system and discover the unforgettable flora, fauna and Amerindian history so special to this tropical paradise. With an unparalleled abundance of wildlife, the Iwokrama Forest is an extraordinary destination for naturalists, birders and for those seeking an authentic retreat to the jungle.

## Resource 9: Iwokrama River Lodge and Research Centre

The River Lodge and Research Centre, situated on the west bank of the Essequibo river, is Iwokrama's operational hub. Although remote, it offers all that you could possibly need for a comfortable stay in the forest.

There are eight beautifully situated river facing cabins, each of which is spacious and beautifully designed, equipped with fans, bathroom, 24-hour electricity supplied by solar power and a wrap-around verandah with hammocks. Our cabins are designed for two but with room for a third bed if required. From these comfortable cabins, watch the sun go down, listen to the many local birds and other wildlife or simply relax in your hammock. These cabins were built for comfort and are perfect for everyone, from families to friends travelling together.

Nearby, overlooking the river, is the bar and the restaurant, where, after an exciting day of discovery and exploration, you can enjoy traditional Guyanese cuisine at our restaurant. A few steps away is a business centre with internet access, a gift shop and a shop selling cold drinks, snacks and general items such as toiletries that you may have left at home. Not far away are two science laboratories and a modern conference room with air conditioning which is wired for digital equipment such as computers and LCD projectors.

## Resource 10: What animals are in the Iwokrama Forest?

The Iwokrama Forest's ecosystem is located at a crossroads between Amazonian and Guianan flora and fauna. Due to this unique location, Iwokrama posesses an incredible animal diversity with an estimated 200 mammals, 500 birds, 420 fish and 150 species of amphibians and reptiles. This includes:

- the world's highest numbers of fish (420) and bat (90) species for any area of this size
- the world's largest scaled freshwater fish – Arapaima
- the world's largest alligator/caiman – Black Caiman
- the world's largest anteater – Giant Anteater
- the world's largest otter – Giant Otter
- the world's largest freshwater turtle – Giant River Turtle
- the world's largest snake – Anaconda
- the world's largest pit viper – Bushmaster
- the world's largest rodent – Capybara
- Americas' largest eagle – Harpy Eagle
- South America's largest cat – Jaguar.

Resources 8, 9 and 10 are quotes from Iwokrama International Centre for Rainforest Conservation and Development website www.iwokrama.org.

# Index

Key terms are shown in **blue type**

# Acknowledgements

The author and publisher would like to thank the following for permission to reproduce material:

Text:

P65 Extract from 'Cost of deluge will be millions' from the Sheffield Star Special Edition. © 2007 by Johnston Press. Reprinted with permission from The Editor of The Sheffield Star, www.thestar.co.uk; p65 Quote from BBC news article ' Why Bangladesh floods are so bad' by Tracey Logan, 27 July, 2004. © bbc.co.uk/news; p108 Extract from BBC news article 'Has China's one-child policy worked?', 20 September 2007. © bbc.co.uk/news; p112 Extract from ' The great betrayal: how the NHS fails the elderly', by Jeremy Laurance. The Independent, 27 March 2006. © 2006 Newspaper Publishing plc. Reprinted by permission of The Independent; p133 extract from BBC News article 'Nairobi slum life: Into Kibera' by Andrew Harding, 4 October 2002. © bbc.co.uk/news; p160 Quote from Gareth Thomas, Minister for Trade and Development, taken from 'How the myth of food miles hurts the planet', by Robin McKie & Caroline Davies, The Observer, 23 March 2008. © 2008 by Guardian News & Media Ltd 2008. Reprinted with permission; p162 Quote from Gordon Ramsay, taken from 'GordonRamsay's war on out-of-season vegetables', by Caroline Gammell, 9 May 2008. © 2008 by Telegraph Media Group Limited. Reprinted with permission; p224 www.thisiscornwall.co.uk

Photographs courtesy of:

1.3A Lonely Planet Images/Woods Wheatcroft; 1.3B Christian Ostrosky/Alamy; 1.3C Frans Lanting/Corbis; 1.3E AFP/Getty Images; 1.4C Pirka-makırı/Getty Images; 1.4D iStockphoto; 1.5C AFP/Getty Images; 1.5I George Steinmetz/Science Photo Library; 1.5J USGS; 1.6C istockphoto; 1.7D Craig Greenhill/Newspix/Rex Features; 1.8B AFP/Getty Images; 1.8C Noboru Hashimoto/Sygma/Corbis; 1.8E Claudia Dewald/Getty; 1.8F Pascal Deloche/Getty; 1.8G Charles E. Rotkin/Corbis; 1.9B Aflo/Rex Features; 2.2C Tony Waltham/geophotos; 2.2G geogphotos film/Alamy; 2.2H(1) © tim gartside usa america/Alamy; 2.2H(2) Photolibrary; 2.2H(3), (4) iStockphoto; 2.3A Tony T/Alamy; 2.3B Jon Sparks/Alamy; 2.4B NIGEL DICKINSON/Robert Harding World Imagery/Still Pictures; 2.4C Reuters/CORBIS; 2.5B Rhett A. Butler/mongabay.com; 2.5C Alex Hipkiss/RSPB; 2.6A imagebroker/Alamy; 2.6C Biosphoto/Eichaker Xavier/Robert Harding World Imagery/Still Pictures; 2.6E Getty Images; 3.1A, C, D Judith Canavan; 3.2B iStockphoto; 3.2F geogphotos film/Alamy; 3.2G iStockphoto; 3.2I Sue Ogrocki/AP/Press Association Images; 3.3F Tony Waltham/geophotos; 3.3G Ed Maynard/Alamy; 3.5B Anna Gowthorpe/PA Archive/Press Association Images; 3.5C Kirsty Wigglesworth/AP/Press Association Images; 3.5E petersmith.com/Environment Agency; 3.5H AFP/Getty Images; 3.6C Xinhua Press/Corbis; 4.1A Gemunu Amarasinghe/AP/Press Association Images; 4.1C Steve Allen Travel Photography/Alamy; 4.2B Jason Hawkes; 4.2C Peter Smith Photography; 4.3A Tony Waltham/geophotos; 4.4B Simon Ross; 4.5A Tony Waltham/geophotos; 4.5B David Lyons/Alamy; 4.5D J Farmar/Skyscan; 4.6B iStockphoto; 4.7A Ian West Photographs; 4.8A Robert Harding Picture Library Ltd/Alamy; 4.8B Manor Photography/Alamy ; 4.8E Martin Keene/PA Archive/Press Association Images; 4.8G Flight Images LLP/Getty Images; 4.9C Peter Hutchings, Hampshire & Isle of Wight Wildlife Trust; 5.2B Threshing the wheat(colour litho) by Tunnicliffe, Charles Frederick(1901-79) Private Collection/The Stapleton Collection/The Bridgeman Art Library; 5.3B iStockphoto; 5.3C Hornbil Images/Alamy; 5.4C NIR ELIAS/Reuters; 5.4D Gareth Brown/Corbis; 5.5B Mangiwau/Getty Images; 5.6A, B, D iStockphoto; 5.7A iStockphoto; 5.7F David Pearson/Rex Features; 5.7G Irene Abdou/Alamy; 6.1C(a) Philippe Bourseiller/Getty Images; 6.1C(b) Fabienne Fossez/Alamy; 6.2A(a) Paul White, Leeds the modern city/Alamy; 6.2A(b), (c), (d) Simon Ross; 6.3A(a) Christopher Pillitz/Alamy; 6.3A(b) Getty Images; 6.3C(a) Judith Canavan; 6.3C(b) iStockphoto; 6.3C(c) iStockphoto; 6.3C(d) Judith Canavan; 6.3D De Agostini/Photoshot; 6.3E(c) Fotolia; 6.3E(d) 2007 Getty Images; 6.3E(e) Fotolia; 6.3F David Flett; 6.3G Judith Canavan; 6.3H Urban Splash http://www.urbansplash.co.uk; 6.4A(a) UPPA/Photoshot; 6.4A(b) Jenny Matthews/Alamy; 6.4A(c) Mark Edwards/Robert Harding World Imagery/Still Pictures; 6.4B Peter Treanor/Alamy; 6.4D Sean Sprague/Robert Harding World Imagery/Still Pictures; 6.4E iStockphoto; 6.5A Getty Images; 6.5B Sipa Press/Rex Features; 6.5C Getty Images; 6.5D Amit Bhargava/Corbis; 6.6A David R. Frazier Photolibrary, Inc./Alamy; 6.6B Titanic Quarter Ltd/Handout/Reuters/Corbis; 6.6C(a), (c), (d) iStockphoto; 6.6C(b) Paul Thompson Images/Alamy; 6.6E UIG via Getty Images; 6.6F Robert Brook/Alamy; 6.6G WilliamRobinson/Alamy; 6.6I Jeff Morgan 12/Alamy; 7.1A Sports Illustrated/Getty Images; 7.1C Sports Illustrated/Getty Images; 7.2B Fotolia; 7.4B Amanda Hall/Robert Harding World Imagery/Corbis; 7.4C scott stulberg/Getty Images; 7.4D PhotoLink/Getty Images; 7.5C Peer Grimm/epa/Corbis; 7.6A Getty Images; 7.7B World Illustrated/Photoshot; 7.7C Getty Images; 7.7D 2006 Per-Anders Pettersson/Getty Images; 7.7F iStockphoto; 8.1B, C, D, E, F iStockphoto; 8.2C Arthur Tilley/Getty Images; 8.3B Evening Standard/Getty Images; 8.3C Chris Ison/PA Archive/Press Association Images; 8.3D Simon Ross; 8.3F Topham picturepoint; 8.3G Rex Features; 8.4B Eli Pascall-Willis/Getty Images; 8.4C Annie Griffiths Belt/Corbis; 8.4D curved-light/Alamy; 8.4F Jason Friend/Alamy; 8.4G World Pictures/Alamy; 8.4H Peter Titmuss/Alamy; 8.4I darryl gill/Alamy; 8.5A Lee Christensen/Getty; 8.5B, F iStockphoto; 8.6A Nordicphotos/Alamy; 8.6B frans lemmens/Alamy; 8.6C Bruce Farnsworth/Alamy; 8.6D, G iStockphoto; 8.6I Arcticphoto/Alamy; 8.7A Martin Shields/Alamy; 10H Eurostat; 10N © Copyright SASI Group(University of Sheffield) and Mark Newman(University of Michigan); 12.1A Owen Braines; 12.1B, 12.1C iStockphoto; 12.2 A(1) Robert Harding Picture Library Ltd/Alamy; 12.2 A(2) David Lyons/Alamy; 12.2C iStockphoto; 12.2D Kenneth Pye Associates; 12.2 F(2) A.P.S.(UK); 12.2 G(2) Robina Herrington; 12.3B Digest of UK Energy Statistics; 12.3H Cornwall Council; 12.3 iStockphoto except 12.3(4) Fotolia; 13.1 qstn 4b John Farmar, Ecoscene, Corbis; 13.3 qstn 1g Photolibrary; 13.3C, D, E Pete Oxford/Minden Pictures/FLPA; 13.3G Iwokrama International Centre for Rain Forest Conservation and Development.

Ordnance Survey maps (2.3C, 3.1F, 3.2C, 3.2E, 3.2J, 3.5F, 6.2B, 9H, 11A, 11C, 11D, 12.3 (7), 13.3 qstn 1) reproduced by permission of Ordnance Survey on behalf of HMSO. © Crown copyright 2013. All rights reserved. Ordnance Survey Licence number 100017284